"In the library I felt better, words you could trust and look at till you understood them, they couldn't change half way through a sentence like people, so it was easier to spot a lie."

Jeanette Winterson

Oranges Are Not the Only Fruit, 1985

Lying: Made in Evolution

LYING
Made in Evolution
Perspectives from
Science and Philosophy

Acknowledgements

I would like to express my gratitude to the authors: Gordon Johnson, Mohamed Hassan, John Hain, Merio and the website PNGdrive.com for the images that I have used in this book.

My thanks also go to Pixabay for allowing access to their online album of free images.

I also would like to acknowledge my enormous debt to all the authors from whose works I have quoted. I feel so very much richer as a result of their valuable works. This book would not have been possible had it not been for their priceless contributions. The last three years of my reading for this project have been amongst the most productive years of my academic life. I thank them all.

If there has been any failure in my interpretation of their texts, or any shortcoming in their citation, I ask their forgiveness.

Allan Nanva

ISBN: 978-0-6487666-0-5 (Paperback)

A catalogue record for this work is available from the National Library of Australia

First published 2020
Designed by Sandra Nanva
isandrananva@gmail.com
+61409696776

Printed by: ARROW PRINT
Jamison town, NSW, 2750, Australia

Contents

Foreword 8

Chapter One
Pinocchio the Polygraph 19

Chapter Two
Biological Rudiments 39

Chapter Three
Psychological Rudiments 59

Chapter Four
Linguistic Rudiments 79

Chapter Five
Mythical & Philosophical Rudiments 95

Chapter Six
Moral and Ethical Rudiments. 131

Endnotes 165

References 173

Foreword

When I set out to write this book I had the ambitious plan of writing a comprehensive book that would cover many aspects of deceptive behaviour, more specifically all modes of lying. The plan was to have the book in two sections: one dedicated wholly to matters evolutionary – part of which makes up this book; the second section was to be solely devoted to manifestations of lying in all its modes and in the context of everyday life-matters: romance, religion, politics, law (eg. false confessions), business, sociability and social linguistics, self-aggrandizement, phatic and other functions of language.

Soon I realized that undertaking such a task was well beyond the ability of one person.

On a revised plan I settled on the idea of limiting the scope of research to the scientific and philosophical groundwork of deceptive behaviour in plants, animals and humans.

Even this plan was in need of tremendous trimming. The scope would still be immensely vast.

I came to the conclusion, eventually, that I needed to be highly selective of topics and relevant theories.

The authorities whose theories have been discussed in this book are prominent and mostly pioneering in their fields. I have tried to limit my references to their work.

I consider my book introductory – introductory in its scope but also introductory in an effort to present complex thoughts and theories succinctly with the intention of giving basic lead to those who wish to pursue the issues more widely or more deeply.

More importantly, introductory by way of making the concepts briefly and simply, as much as possible, to be accessible to the common reader.

Finally my hope is that the book distinguishes itself, by virtue of its blend of scientific and philosophical views, albeit at the threshold level, from hundreds, perhaps thousands of materials published in various forms on the subject of lying.

Allan Nanva

"The Fox and The Crow

A crow who had stolen a piece of cheese was flying toward the top of a tall tree where he hoped to enjoy his prize, when a fox spied him. "If I plan this right," said the fox to himself, "I shall have cheese for supper."

So, as he sat under the tree, he began to speak in his politest tones: "Good day, mistress crow, how well you are looking today! How glossy your wings, and your breast is the breast of an eagle. And your claws-I beg your pardon-your talons are as strong as steel. I have not heard your voice, but I am certan that it must surpass that of any bid just as your beauty does."

The vain crow was pleased by all the flattery. She believed every word of it and wagged her tail and flapped her wings to show her pleasure. She liked especially what friend fox said about her voice, for she had sometimes been told that her caw was a bit rusty. So, chuckling to think how she was going to surprise the fox with her most beautiful caw, she opened wide her mouth.

Down dropped the piece of cheese! The wily fox snatched it before it touched the ground, and as he walked away, licking his chops, he offered these words of advice to the silly crow: "The next time someone praises your beauty be sure to hold your tongue."

Aesop's Fables

Magnum Books, USA, 1968, PP.11-12

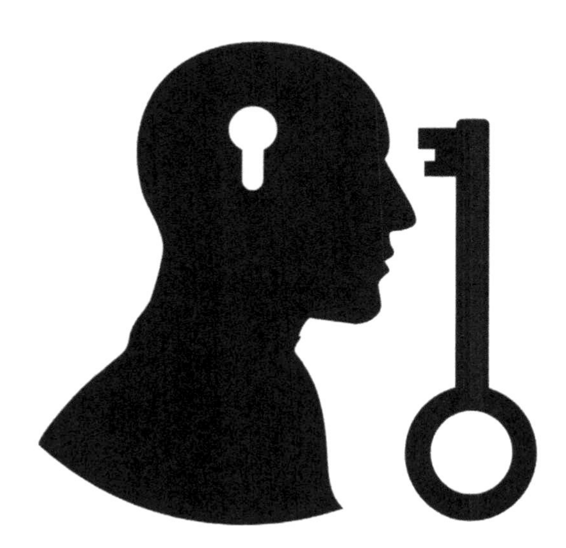

Deception is driven by intelligence.

Intelligence is an aspect of cognition of which problem solving is a feature. Problem solving involves thinking of new or different ways to overcome a challenge. Microscopic organisms, insects, birds, animals and humans are equipped with this ability. Deception employs this ability.

DARWIN

Lying is a mode of deceptive behaviour peculiar to humans. Deceptive behaviour for self-preservation is not peculiar to humans. Deceptive behaviour is an evolutionary trait that exists in all organisms - from microscopic such as bacteria, to plants, animals and, of course, humans. The phenomenon of deception, why and how it happens, has fascinated philosophers, theologians as well as scientists for centuries.

This book looks at the phenomenon of deception, lying in particular, from various perspectives: biology, psychology, linguistics and philosophy, albeit at rudimentary level.

"Evolutionary biology teaches us that the tendency to deceive has an ancient pedigree."

David Livingstone Smith
WHY WE LIE

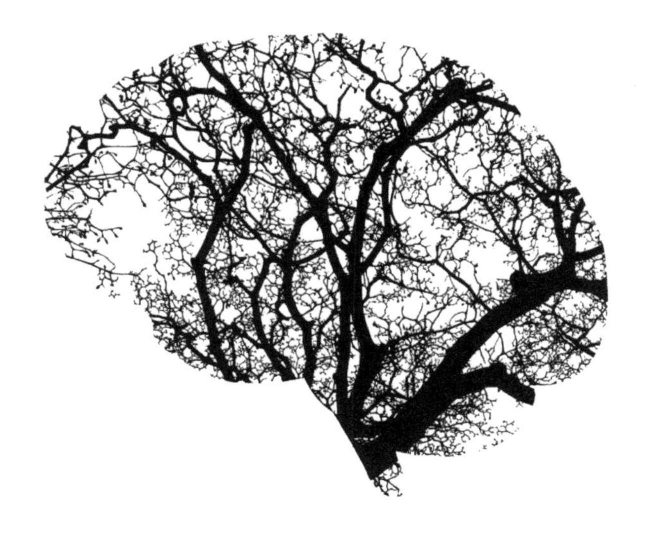

"If you have dealt with liars, even pathological ones who pass polygraph tests, you know the signs to look for. The liar blinks just before the end of the lie, or he keeps his eyelids stitched to his brow. He folds his arms on his chest, subconsciously concealing his deception. The voice becomes warm, a bit saccharine; sometimes there's an ethereal glow in the face. He repeats his statements unnecessarily and peppers his speech with adverbs and hyperbole. The first-person pronouns 'I,' 'me,' 'mine,' and 'myself,' dominate his rhetoric.

Conversely, the truth teller is laconic and seems bored with the discussion, not caring whether you get it right or not."

James Lee Burke

Robicheaux, 2018

Chapter One

Pinocchio the Polygraph

Exaggeration is a kind of lying.

"The Lying Lion

He lied, oh how he lied,
Claiming to be king
But in actuality, this lion cried,
At the thought of a little sting.

I witnessed it myself that day,
And couldn't believe my eyes;
As this big bad cat began to sway,
Desparate to conceal his lies.

He ran away as fast as he could,
Hoping to get lost;
Feeling embarrassment and misunderstood,
Came at quite a cost.

It had been easy for him to scare us,
With his mighty roar and size;
But to see him tremble and throw a fuss,
Felt like winning a first prize.

I knew someone wiuld eventually call his bluff,
And that they would reveal the truth;
I just never would have thought, all that fluff,
Would be caused by a bee named Ruth. "

Artie Knapp Stories
dltk.teach.com

"Cathy's lies were never innocent. Their purpose was to escape punishment, or work, or responsibility, and they were used for profit. Most liars are tripped up either because they forget what they have told or because the lie is suddenly faced with an incontrovertible truth. But Cathy did not forget her lies, and she developed the most effective method of lying. She stayed close enough to the truth so that one could never be sure. She knew two other methods also – either to interlard her lies with truth or to tell a truth as though it were a lie."

John Steinbeck
East of Eden, 1952

Pinocchio the Polygraph

> 'Have you heard the great news?'
> 'No.'
> 'In the sea near here a Shark has appeared as big as a mountain.'
>
> Pinocchio by Carlo Collodi

The story of Pinocchio, the 19th century novel by the Italian writer Carlo Collodi /Lorenzini (1826-1890), teaches us three things in relation to lying: that we are born with the ability to lie; that we grow to like lying; and that Pinocchio's nose is a physical manifestation of lying.

More importantly, Pinocchio's nose is the 19th century equivalent of the early 20th century lie detecting machine, the polygraph (invented by John A. Larson in 1921) and the late 20th century invention, fMRI (Functional Magnetic Resonance Imaging), invented in Bell Laboratories, USA, by a group of researchers under Seiji Ogawa in1990).

Collodi's fairy tale, a fantasy novel - The Adventures of Pinocchio (1883), begins when the puppeteer Geppetto receives, as present, a strange piece of wood from his carpenter friend, Master Cherry. The Piece of wood laughs and cries like a child, so it is just the right piece of wood for Geppeto the puppeteer who has been on the search for a piece of wood that would make a wonderful puppet.

While busy crafting it, the puppeteer gives the piece of wood the name Pinocchio -

"Having found a name for his puppet he began to work in good earnest, and he first made his hair, then his forehead, and then his eyes. Having done his eyes, just think of his astonishment when he noticed that they could move and were looking straight at him. Geppetto, finding himself stared at by those two wooden eyes, felt almost offended and said angrily 'You naughty wooden eyes, why are you looking at me?' No one answered. He then proceeded to carve the nose; but no sooner had he made it than it began to grow. And it grew, and grew, and grew, until in a few minutes it had become an enormously long nose that seemed as if it would never end. Poor Geppetto tired himself out with cutting it off; but the more he cut and shortened it the longer did that impertinent nose become". [(1)]

The puppeteer continues crafting the wood by carving a mouth into it and the mischief begins. Pinocchio sticks out *his* tongue - *principal means of verbal lying* - at the puppeteer, steals his wig and runs away leaving the man in awe.

Back at home after a series of adventures and misadventures, a hundred year (or so) old cricket on the wall accosts Pinocchio, advising him against consequences of children disobeying their parents. The angry Pinocchio hammers the wise old cricket to death.

When Geppetto returns home the next day, he finds that Pinocchio has burned his feet.

Pinocchio pleads to the master puppeteer to give him new feet. He promises to be a good boy and pledges to go to school to get educated if only the master puppeteer gave him new feet.

Not only does the puppeteer grant Pinocchio his wish, he, believing in the value of education as a means to a truthfully rewarding life, sells his winter coat in the market to buy Pinocchio his school book.

Geppetto ignores a number of Pinocchian inaccuracies, thinking children do, *naturally,* lie. He himself lies to Pinocchio saying he had sold his coat because he was feeling too hot (social lying). This is the first outright lie in the book - *adults as well as children lie for whatever reason.*

As it happens with liars, Pinocchio forgets his promise when he follows the sound of pipe music to a puppet theatre, sells his spelling book that Geppetto had bought for him to purchase an entry ticket - *distractions, as well as time itself, are the enemy of liars; that is, us.*

Once inside, Pinocchio is taken onto the stage by other puppets to perform. The Fire Eater is badly annoyed. He tells the puppet Pinocchio that as punishment it is going to be used for fire wood to cook dinner.

This time Pinocchio is overtaken by fear - *one of the principal causes of telling lies both in children and in adults. As a major reason for lying, fear is also a key trigger of brain response to lying recorded on fMRI images.*

The Fire Eater feels pity and agrees to use another puppet in his place. The Fire Eater's decision saddens Pinocchio. He tells the Fire Eater, if so, he would jump into the fire pit to save the other puppet. The Fire Eater is touched by Pinocchio's honesty and rewards him with five gold pieces - *truthfulness and honesty win rewards.*

On the way home, Pinocchio encounters a lame fox (feigning) and a blind cat (feigning), tricking him into burying his gold pieces in the Field of Miracles where the five would grow to 2000 gold pieces. In the field he is warned by the ghost of the wise cricket he had killed not to believe the deceptive fox and cat. Pinocchio ignores the advice and goes ahead with the plan - *cupidity is another motivating cause of lying.*

In the meantime another misadventure befalls Pinocchio - he is hung on an oak tree by two assassins but, almost near death, very sick, he is saved by the gracious Fairy who quizzes Pinocchio on the whereabouts of the gold pieces-

"I have lost them!' said Pinocchio; but he was telling a lie, for he had them in his pocket. He had scarcely told the lie when his nose, which was already long, grew at once two inches longer.

'And where did you lose them? '

'In the wood near here.' At this second lie his nose went on growing.

'If you have lost them in the wood near here,' said the Fairy, 'we will look for them, and we shall find them; because everything that is lost in that wood is always found.'

'Ah! now I remember all about it,' replied the puppet, getting quite confused; 'I didn't lose the gold pieces, I swallowed them by accident whilst I was drinking your medicine.'

At this third lie his nose grew to such an extraordinary length that poor Pinocchio could not move in any direction. If he turned to one side he struck his nose against the bed or the window-pane, if he turned to the other he struck it against the walls or the door, if he raised his head a little he ran the risk of sticking it into one of the Fairy's eyes. And the Fairy looked at him and laughed.

'What are you laughing at?' asked the puppet, very confused and anxious at finding his nose growing so prodigiously.

'I am laughing at the lie you have told.'

'And how can you possibly know that I have told a lie?'

'Lies, my dear boy, are found out immediately, because they are of two sorts. There are lies that have short legs, and lies that have long noses. Your lie, as it happens, is one of those that have a long nose.'

Pinocchio, not knowing where to hide himself for shame, tried to run out of the room; but he did not succeed, for his nose had increased so much that it could no longer pass through the door." (2)

Having recovered from illness and embarrassment, Pinocchio returns to town and takes the road to the Field of Miracles, imagining himself very rich. At the field and looking, Pinocchio hears the explosion of laughter by the parrot:

"Why are you laughing?' asked Pinocchio in an angry voice.

'I am laughing because in the pruning my feathers I tickled myself under my wings." (3)

Of course the Parrot was lying. Later the Parrot told Pinocchio the truth – the fox and the cat had taken the gold pieces.

Upon reporting the theft to the 'courts of justice' Pinocchio, to his utmost surprise, is jailed for being so gullible in the society. On his release Pinocchio starts searching for his 'father' who has gone searching for his 'son'.

In the Land of Toys Pinocchio and his friend Candlewick are turned into donkeys. A musician buys the donkey Pinocchio to use his hide to make a drum. Thrown into the sea with a rock tied around its neck, the donkey is changed back into a puppet by a fairy but gets swallowed by a shark. He finds Geppetto in the shark's belly from where they escape. With the help of the fairy, Pinocchio is morphed into a real boy. He works hard in the nearby farm, applies himself to studying at night and lives happily with the old master Geppetto in a bright home from then on.

***

The subject of deceptive behaviour has been of interest to scientists, to philosophers, composers of works of literature and art for centuries (as we shall see in the coming chapters).

With the advent of the internet and new developments in brain imaging technology, research, at different levels of scientific sophistication, has become more than ever possible and accessible. The pace of research in the area has increased dramatically not only because of the availability of tools of research and dissemination of information worldwide, but also because of the increased, and increasing, appetite for information in all matters of life (and death) by all people.

Complexities of modern life amongst ever growing world population competing for all sorts of satisfaction – emotional, educational, food, jobs, shelter, vital natural resources (water, for instance), health services, social and political power, most particularly economic reasons, have made deception one major sociological issue - all that augmented, aggravated, by the grand, convenient lies of political leaders who, often in collusion with, or under the influence of, powerful bodies and big corporations, engage in extraordinary manipulation of truth: cover ups, state sponsored sabotaging of facts, faking realities on or off line, using religious faith and patriotic sentiments to substitute facts, abusing international platforms such as the United Nations to justify their international crimes, camouflaging weapons of mass destruction in the shape of gases, microbes, chemicals, plutonium radiation. This list is by no means exhausted but people's trust in their institutions such as political and religious (etc.) is.

Untruth for truth is beamed down to us non-stop via satellites from where we thought, not too long ago, was Heaven where god - The Truth - resided.

Further, the growing number of graduates in the fields of 'pure' and 'social' sciences are finding niche topics to engage with.

Now that traditional safeguards against deceptive behaviour - things like strong religious, moral and ethical convictions - scarcely exist in most societies (if they ever truly did), and where they do now it is predominantly for convenience and opportunism, the growth of research into the causes and impacts of deceptive behaviour is a natural academic development.

We witness daily the alarming exposé of lies and sophisticated deception committed by people and institutions that we had learned to trust: the legal and law enforcing system, national security and defence, welfare areas, banking and financial sector - areas that greatly impact lives of individuals and the affairs of the state.

Given such developments, the growing demand for more accurate, technologically reliable means of detecting lies – as accurate and as physically manifest as Pinocchio's nose, is only *natural.*

In an effort to detect lies, researchers have, for some time now, been aided by technology - the polygraph, the machine that replaced Pinocchio's nose. Given the inaccuracies associated with that technology, research has now turned to the newer science of Functional Magnetic Resonance Imaging or the fMRI machine.

Still existing and sometimes used by law enforcing agencies amongst others, the polygraph is a machine that measures the stress caused by anxiety, in this case elevated by lying. The elevated stress leads to increased heart rate, more rapid breathing than normal, elevated blood pressure, perspiration and other physical symptoms that get matched against false responses to certain questions, particularly questions that stimulate those symptomatic physiological reactions.

The reliability of the polygraph is in doubt for various reasons. For example, if the subject, eg., a psychopathic liar, is unusually calm during the test, the result can be affected by a considerable variation. Similar variation can occur if the subject is gravely anxious even though he/she has nothing to hide except that he/she has been subjected to the test on grounds of suspicion. As a result of these inaccuracies, the newer fMRI technology is being used (work-in-progress) to eventually replace polygraph testing.

The fMRI machine, on the other hand, reflects magnetic signals from oxygen atoms attached to iron molecules in the person's blood. An increase in brain activity is followed by increase in blood flow to the brain, carrying extra oxygen with it causing signal strength. The machine records, maps rather, activity in millimeter-sized parts of the brain. Experiments have shown that lying triggers such nuances prominently in sections of the brain known as the frontal, temporal and limbic areas.

Research, utilizing the fMRI technology to detect lying, is extensive and vary depending on the methodology used but, by and large, they point to the same results.

Generally, indications are that when we lie, in contrast to when we tell the truth, a number of actions and reactions occur in the brain which the fMRI machine represents as images on devices for experts to interpret.

Repeating data here would be redundant mainly because they are accessible on the internet, but also because my intention is to avoid transgressing the introductory nature of the book.

Summarily speaking, lying as a cognitive function in humans is for the purposes of concealing (oneself) - hiding oneself from oneself (telling lies to oneself), hiding oneself from others, revealing oneself in shades of the truth like showing off the wealth that one does not have, wearing brand clothes or driving brand cars, wearing make-up or making up stories about one's personal background, about one's beliefs, one's interests, one's family, one's relationships and the like.

Lying begins from the moment one begins to talk, if not earlier, and dominates one's talking throughout life. It is the constant variable in human sociability.

More broadly, deceptive behaviour is an innate ability in all organisms for the purpose of survival. Plants, insects, animal and humans are all beneficiaries of the ability to deceive. Deceptive tendencies simply exist in nature; and in humans, like in other organisms, nature manifests itself in strange ways to which Pinocchio's nose bears testimony.

Similar to Pinocchio's nose, the fMRI machines have shown lying to have triggered activation of a number of cognitive control processes namely the frontal and parietal cortices, as well as in cortices responsible for evaluating social behaviour namely the superior temporal cortex and temporal poles.

When we are challenged by a circumstance in which we do not wish to reveal the truth, whether for social or anti-social reasons, we begin to talk to ourselves silently, thinking (thinking is the act of entering a task dependent state) how to avoid, manipulate, hide (by means of verbal/nonverbal language) the truth which our brain already knows.

While this inner talking (thinking, deciding) is in progress (hence the increase in neural activities), the process engages a number of the brain's higher cognitive functions: our frontal lobe (brain's personality region), the temporal lobes on each side of the brain (memory and emotion regions) as well as the limbic system.(for positive, negative emotions and self-preservation region) - activities that the fMRI can detect and map out.

Of particular interest to research is the function of the two Amygdalae in the temporal lobe. The word Amygdala is a Greek word meaning 'almond'. The name is applied to the brain organ because of its almond shape. The Amygdale enable us to feel emotions and to perceive emotion in other people; for instance, if we take the case of fear - fear is one necessary emotion that warns us of imminent danger in response to which the two Amygdalae modulate our actions.

In most cases the definition of lying was, as it should be, based on the premise that the person must know the truth which she/he decides, spontaneously in the case of laboratory studies, to substitute with something else. Further, that the subject must remember (inner self) the specific details of the experience about which she/he is quizzed or placed under study/investigation, taking into account the subject's opinion, beliefs and emotions about the matter at hand. The study/investigation is to be modeled to involve cognitive control of the way the subject might inhibit or manipulate the true information.

It is a scientific premise that to put together a lie our brain collects information relevant to the experience from memory regions of the temporal lobes – declarative memory and/or episodic memory - passes on the information to the frontal cortex which gives the person the chance to decide whether or not to suppress what the brain knows to be true.

Declarative memory is defined as one that holds facts and information related to previously experienced events. The retrieval of declarative memory is consciously executed.

Episodic memory, on the other hand, is the recall of emotional and sensory experience of events. Recall of episodic memory is involuntary and is triggered by stimuli, the smell of a specific perfume, for example.

The right Amygdala is thought to have the role of associating time and place to the emotional experience.

Naturally any detection of lies related to past experiences involves sections of the brain that support maintenance/retrieval of memories of the episode - regions like the medial temporal lobes and the prefrontal cortex.

Lie detection needs to identify a number of neural correlates and variables: whether the person chooses to lie about their personal views based on social pressures, the norms of the people around, cost vs. benefit analysis and, ultimately, the effects on their relationships in whatever context.

The regions of the brain associated with moral and ethical reasoning are known to include frontal and parietal, medial frontal cortices, the superior temporal sulcus and the temporal parietal junction.

Despite its apparent simplicity, lying, rather deceptive behaviour generally, is multi-layered and complex. Like other cognitive skills, the layers are formed over time depending on the physical, psychological and sociological environment in which the person is born, in which the person grows up and to which the person learns to adapt in order to survive.

"Food-
Crouched in grasses - sun is sinking
food will come for evening drinking
silent - still - panting - thinking,
resting in the heat of sun.

Sun goes down and sky is red
food is coming - scent is read
lust for blood engulfs my head,
muscles tense for coming run.

Here they come by twos and threes
down to water - on their knees
no scent of me is on the breeze,
my hunt is only now begun.

There's one limping - limping still
he can't run with speed or skill
he'll be my food - my easy kill,
mark him well for he's the one.

Rising - tensing for the fray
alarmed - the herd stampedes away
thundering hooves - all but my prey,
my hunger peaks - this hunt is won.

Explode from cover - extending claws
closing fast on lightning paws
pounce - his throat between my jaws,
hold him down till kicking's done.

Feasting now in cool of night
flesh tastes rich and blood is bright
hyenas hanging back in fright,
I'll hunt again with newborn sun."

Ted L. Glines
PoemHunter.com, 2009

Chapter Two

Biological Rudiments

"Animals, even plants, lie to each other all the time, and we could restrict the research to them, putting off the real truth about ourselves for the several centuries we need to catch our breath. What is it that enables certain flowers to resemble nubile insects, or opossums to play dead, or female fire flies to change the code of their flashes in order to attract, and then eat, males of a different species?"

Lewis Thomas

Late Night Thoughts on Listening to Mahler's Ninth Symphony (1984)

Lying is natural.

Deceptive skills are predatory as they are predator-avoidance mechanisms in organisms. They are natural skills necessary for the survival of genes to continue producing and reproducing generations of themselves.

"Cats are opportunistic predators by nature. If given a chance to kill a bird or other small animal, most cats will take it. That is just the way cats are made."

Peter P. Marra

Cat Wars: The Devastating Consequences of a Cuddly Killer, 2016

Biological Rudiments

Lying is natural.

Natural not in the sense that it is universally common. That is given.

Natural in the sense that it is an evolutionary phenomenon - it exists in nature; therefore, it is a biological requirement of survival in the competitive world of organisms: plants, animals and, of course, humans.

The history of cheating and deception on earth began some two billion years ago when single cell organisms evolved into multicellular colonies through complex evolutionary processes.

About a billion years ago, multicellular organisms generated societies and in them grew tissue differentiation and inter-cellular communication. As these cellular societies grew and spawned, so did the need for organization and regulation of the member cells. "The result?", ask researchers Athena Aktipis and Carlo C. Maley in a ground breaking article titled 'Cooperation and Cheating as Innovation: Insight from Cellular Societies' (2017).

Aktipis and Maley hold the view that multicellular bodies demand both cooperation and the management of cheating and conflict through implicit rules that are coded into the genomes of each cell. These rules include both directions for how to cooperate and restrictions on cellular behaviour that keep cells from cheating and exploiting the organism.

The authors suggest that these multicellular societies are comparable to human societies living together and interacting in a community.

Many multicellular societies promote the common good, prevent exploitation and help coordinate higher level functions like resource transport and division of labour; thus, every multicellular organism can be considered a society of cells, regulated not by explicit laws, rather by rules encoded in genomes -

"The tension between cooperation (which benefits the organism/group) and cheating (which benefits the individual) is central to the evolution of multicellularity.... Multicellularity represents a suite of innovations for cellular cooperation, but [it] also opened up novel opportunities for cells to cheat, exploiting the infrastructure and resources of the body". (4)

One way of preventing cheaters from exploiting vulnerable cells, according to Aktipis et al, was to "deploy additional innovations including the adaptive immune system and the development of partnerships with preferred microbial partners". [5]

Cooperation amongst cells in the genetic networks is vital for maintaining cell fitness and health as well as for suppressing cellular cheating that happen - for example, cancer cells or viruses or bacteria that cause the common cold. Aktipis et al define cheating as, "simply the breakdown of shared rules (including the genetically encoded phenotypes or behaviors) that leads to fitness advantage on the cellular level for the cheater" [6].

It was not until Charles Darwin's theory of evolution, particularly his views on the natural selection of species, [7] and his other research, that science assumed its rightful role in demystifying deceptive behaviour as an aspect of evolution amongst all species, not the least amongst humans.

Darwin suggested that there was a genetic cause behind lying (by organisms) - and in the case of humans, fundamental to and alongside other factors such as sociocultural, psychological and economic, some of the major agents known to drive deceptive behaviour.

In an interesting article in New York Times, under the heading 'Cellular Cheaters Give Rise to Cancer' George Johnson (2015), based on research done by Athena Aktipis (her research in its entirety was published in a book released in March 2020 titled 'The Cheating Cell, New Evolution helps us

Understand and Treat Cancer') wrote -

"Maybe it was in some warm little pond, Charles Darwin speculated in 1871, that life on Earth began. A few simple chemicals sloshed together and formed complex molecules. These, over great stretches of time, joined in various combinations, eventually giving rise to the first living cell: a self-sustaining bag of chemistry capable of dividing and spawning copies of itself. While scientists still debate the specifics, most subscribe to some version of what Darwin suggested – genesis as a fortuitous chemical happenstance.... As the primordial cells mutated and evolved, ruthlessly competing for nutrients, some stumbled upon a different course. They cooperated instead, sharing resources and responsibilities and so giving rise to multicellular creature – plants, animals and eventually us. Each of these collectives is held together by a delicate web of biological compromises. By surrendering some of its autonomy, each cell prospers with the whole. But inevitably, there are cheaters: a cell breaks loose from the interlocking constraints and begins selfishly multiplying and expanding its territory to the free-for-all of Darwin's pond. And so cancer begins". (8)

Cancer cells are amongst the worst deceiving scavenger cells as they confuse (dedifferentiate) (9) the healthy cells to their own advantage.

Our cells normally go about their routine business of replicating themselves no more no less than necessary to sustain life for us. There are times when a cell deviates from the norm for whatever triggering factor, and acts contrary to the behaviour of the colony, in which case our defensive cells are quick to try and destroy them through a mechanism called 'programmed cell death or cellular suicide'. Cancer cells engage seriously in an attempt to defeat this safeguard. 'They refuse to die'. [10]

Scientific literature abounds with research on cheating bacteria. In one case, for instance, a research team working under Karina Xavier from 'Instituto Gulbenkian de Ciencia' in Portugal , found that "Similar to human societies bacteria live in communities and interact with each other in order to better adapt to the environment. Regularly, they produce compounds which behave as public goods, as they are secreted to the surroundings and are consumed by the entire population, benefiting the whole bacterial community. However, some bacteria carry mutations that prevent them from producing those public goods. These bacterial mutants act as cheaters, benefiting from the public goods without contributing to their production". [11]

It needs to be said that Darwin (1809-1882) was not the first person to have discovered the general concept of hereditary nature, deceptive or otherwise. That is attributed to the Czech scientist, Gregor J. Mendel (1822-1884) of whose discovery Darwin is said to have benefited to explain that organisms enjoy heritable characteristics as part of, and essential to, their regeneration and reproduction.

Mendel, now acknowledged as the father of modern genetics, [12] had found, through experiments on pea plants, that the key to the survival success of an organism was its having the advantage of adaptive genes empowering it to succeed against equally desiring but less capable or less adaptive competition.

'Adaptation to what?' [13] Inquires John Cartwright, in his discussion of 'Evolutionary Psychology or Darwinian Anthropology?'

The answer, it seems, is inherent in the organism's own efforts, via trial and error, to adopt the best (fittest) mechanism to survive in a given circumstance. Once the mechanism is successfully adopted and its effectiveness assured, the organism ensures passing the traits to its future generations via what Konrad Lorenz (1903-1989), the famous Austrian ethologist, calls the 'fixed action pattern', the continuous repetition of a given behaviour in a linear line of species stimulated in response to some environmental (external) factors.

Through a repetitive pattern, behaviours become instinctive, demonstrating the following characteristics -

"Their form is constant, that is the same sequence of actions and the same muscles are used. They require no learning. They are characteristics of a species. They cannot be unlearnt. They are released by a stimulus". [14]

Darwin maintained that reproduction is a selection process driven by what he called 'sexual selection'- the struggle of the male of the species to access females, an indispensable aspect of the 'natural selection'. [15]

The phenomenon of sexual selection, according to Darwin, is dependent upon two mechanisms: Intersexual and intrasexual, the latter selection meant competition between members of the same sex (predominantly males) for access to mating partners and the former referred to the members of one sex, (predominantly females) to choose members of the opposite sex.

While Darwin is universally celebrated for his contributions to geological and zoological sciences, his ideas on plant regeneration have been no less valuable to the botanical scientific research. For instance, while investigating evolutionary traits: speciation, mating systems, sensing and tropism of living organisms, "He was surprised to discover that many orchids used fraud and deceit to achieve pollination by insects....". [16]

Darwin's pioneering experiments on plant biology and plant signalling molecules have now reached scientific truism, indicating, says David Livingstone Smith in his book, 'WHY WE LIE' (2004), that, "it is more useful for living things to exploit one another by sending dishonest signals....One option is to use the art of seduction". [17]

Mirror orchids (Orphys speculum), for instance, use deception by sexual signalling to attract pollinators – an act of plant pornography -

"The flower's scheme begins with the release of a strong scent that simulates the pheromones (sexually arousing chemicals) released by female wasps. The artificial wasp scent is actually hyper potent: it is so effective that males are more attracted to the scent of the orchid than they are to that of real females. The intoxicating fragrance, combined with the beguiling shape of the flower, lures the male wasp into an arousing yet ultimately frustrating cross-species erotic entanglement. In a real copulation, males use the hairs on the female's abdomen to get themselves into the right position for mating. The specific stimuli that would bring about ejaculation are not present, so the wasp, unable to obtain satisfaction, lingers on the flower, picking up more and more pollen in his desperate attempt to breed. Biting the flower in frustration, he only succeeds in releasing more imitation wasp-pheromones to bewitch his senses; and when he finally leaves, it is not long before he falls under the seductive spell of another flower, depositing on it some of the pollen that is still clinging to his body, and picking up some more, thus unwittingly serving the orchid's reproductive interests". [18]

We are mesmerised by T.V documentaries on animal-eating plants (carnivorous plants) of which the Venus flytrap (Dionaea muscipula) is a star attraction and one that has acquired the nickname 'Flora fatale'.

Venus flytrap and other insect eating plants have learned to survive in geographical locations where soil nutrition is inadequate. They have adapted by environmental necessity to eating insects and a variety of arthropods such as centipedes, scorpions and beetles.

The Venus Flytrap, a species of Snap Trap, have two trap lobes hinged together at the base looking like two human palms held against each other half open in a convex position. The lobes are equipped with touch sensitive hair which when triggered by a moving prey create a chemical reaction resulting in the lobes to snap shut tightly in less than a second, entrapping the prey to death and digesting it over a period of up to two weeks . In order to avoid false activation of the lobes by rain drops or windblown debris the lobes are equipped with a built-in memory mechanism that requires two time-delayed activation (anything from 0.5 to 30 seconds apart) to respond.

In an Australian Broadcasting Commission (ABC) science report, the reporter Ann Jones documented a carnivorous plant in the Blue Mountains area of Sydney, a species of the Drosera plants commonly known as the Sundews. In her posted article and quoting a Drosera researcher, Jones explains how these plants use their sticky tentacles to attract, then embrace their prey. As the prey struggles to free itself, it comes in contact with some incapacitating gland activation that flicks the prey to the centre of the leaf ready for digestion. Jones writes, "In fact, their lethal speed can be faster than a cheetah at full gallop". [19]

One Australian species 'Drosera glanduligera', the researcher is quoted to have said to Jones,"moves so rapidly it's faster than the eye can see. It's a couple of hundredths of a second". [20] Under the heading 'The plant that frightened Darwin' Jones quotes a statement by Charles Darwin in 1960 as saying 'I care more about Drosera than the origin of all the species'. [21]

The study of sexual selection in animals is considered one of Darwin's major contributions to ethology – the study of animal and human behaviour from a biological perspective.

Darwin's natural selection theories encompass some non-sexual evolutionary strategies that species use to avoid predators and develop physical traits; however, his theories on sexual selection help understanding animal behaviour in the context of attracting the fittest (and the prettiest) sexual partners. The question inevitably arises: how does this selection take place?

According to Darwin it is predominantly the female of the species that is endowed with the choice. The male on the other hand has to show physical strength, vivid adornments and beauty of the body to get chosen by the female, hence the male peacock's extravagantly bright and showy tail or the male lion's flowing neck hair. While these traits are necessary in animal courtship, they may not be sufficient by themselves to attract a female partner. The male may need to use smart ploys and deceptive techniques to complement those favourable physical attributes; It was Darwin who first theorized that these vivid adornments that distinguished male animals from females evolved through sexual selection, the process by which a female, seeking the 'best' sire for her offspring, chooses a male that flaunts the most colourful raiment, or shows the most impressive physique, or gives the appearance of being a good provider, or that has, as Darwin said 'the power to charm the female. [22]

Deceptive strategies are particularly beneficial for larger animals, both male and female, that need to keep their presence secret from their prey before they carry out an ambush. We see crypsis - the ability of animals to avoid detection by using camouflage or acts of visual, olfactory (smell) or auditory mimicry - in our domestic cats how they lie low and flat on their tummy before jumping into action to catch a chirpy bird picking seeds off some plant, unaware. However, deceptive techniques are essential for animals of all shapes and sizes whether to be a successful predator or a successful survivor.

There are many species of animals that morph themselves into deceptive shapes to avoid being detected. Tatu Virando Bola (Armadillo), for instance, rolls itself into a ball utterly indistinguishable as an animal.

Scientists, Paul Weldon and Gordon Burghardt (1985) have found that animals, whether mammals, fish, reptiles, birds or invertebrates are skillful in acts of mimicry and deception.

The two scientists contend that some male animals, for instance, go as far as mimicking members of the opposite sex (transvestism) to avoid being attacked by other male competitors, and approach females without making them suspicious prior to engaging sexually with them. Similarly, some younger individuals of a species adopt the looks and actions of older males deceptively to mate with females that would normally prefer older males for mating. [23]

Darwin reminds us (in The Origin of Species) that such acts of deceptive behaviour are instinctive, "...everyone understands what is meant, when it is said that instinct impels the cuckoo to migrate and to lay eggs in other birds' nests. An action...performed by an animal, more especially by a very young one, without any experience, and when performed by many individuals in the same way, without their knowing for what purpose it is performed". [24] One interpretation of such action by the cuckoo suggests that this is a parasitic act of deception by the cuckoo bird to put their chicks in competition for food with the host's own babies.

Another study by Daniela Canestrari suggested that this was an act of mutual benefit to the cuckoo and the crow chicks as the smell of a secretion from the newly hatched cuckoo chick deterred predators, thus helping both the crow and the cuckoo babies to survive .[25]

Parasitic behaviour to benefit from others by deception is also found in the beetle known as Atemeles pubicollis (Joseph Parke, 2016). In order to enter the colony of ants, secured by the smell of a recognizable chemical substance, the beetle generates the familiar odour of the substance as evidence of belonging to the colony. The beetle gets adopted by the unsuspecting ants and "Once inside, it is fed by the ants of the colony and then begins feeding on the eggs and larvae that were being sheltered in this environment". [26]

Another significant point to be made, given its importance to our future discussions on psychological reasons behind deceptive behaviour, is Darwin's fine line between instincts and habits.

Unlike instinctive behaviour, habitual behaviour has a component of intentionality and can be modified 'by will or reason', although Darwin does not deny their close resemblance, "If we suppose any habitual action to become inherited – and I think it can be shown that this does sometimes happen – then the resemblance between what originally was a habit and an instinct becomes so close as not to be distinguished". [27]

David Barash, in his book, 'Sociobiology: The Whispering Within' (1981), recounts the story of an ocelot - a wild cat - in the Seattle zoo, U.S.A, that plucked its own fur violently, literally ripping patches of skin from its body as if intending to torture itself to death. The worried zoo keepers tried the best they could to help, anything from a bigger cage to a female ocelot companion were thought of and provided, all to no avail. The staff at the zoo had failed to realize that in the wild, ocelots mainly eat birds and they have the habit of plucking feathers off their freshly hunted preys before they make a meal of them. Some good thinking staff member finally decided to give the cat a fully feathered chicken rather than the usual horse meat. The cat started plucking the feathers off the chicken and immediately stopped ripping its own fur. (28)

The issue of intentional vis-à-vis instinctive mimicry in animals, particularly in apes from whom, according to Darwin, humans have descended, has for a long time attracted the attention of primatologists.

There are numerous anecdotes of scientific observations of intentional/instinctive deception carried out by chimpanzees and gibbons, our closest ancient ancestors, that one can find in materials on comparative behaviour of animals and humans.

In one anecdote reported by Cecilia M. Heyes of Oxford University (1994), a female baboon desiring the carcass of an antelope that the male baboon was holding approached him mimicking a caring female and began to groom him. When the male baboon fell for her charm, began to rest and stretched himself in total relaxation under her attention, the female baboon snatched the carcass and ran with it leaving the male baboon in awe. (29)

The term 'anecdote' is appropriately used here as it allows factoring in reports of events that may have happened as a result of circumstantial opportunism and chance driven functional deceptive behaviour. While it is true that actions of one individual in a given circumstance cannot be regarded as inherently typical of a species, scientists (Helmut Sick, 2018) are in agreement that deceptive behaviour in species, eg. acts of camouflage, has an instinctive base albeit manifested differently in different environments by different individuals.

The German/Brazilian ornithologist, Helmut Sick (1910-1991) while investigating hawks behaviour found that a hawk species known as Barred-Tailed or Buteo albonotatus, deceives its unsuspecting preys by flying with members of non-predatory bird species. Like other accompanying birds it flies above its prey until in control of the situation before it jumps onto the oblivious prey.(30)

Unlike animals that can only use chemical and mechanical tactics to carry deceptive information to other animals for food or predatory debarring, humans are able to use language, non-verbal but predominantly verbal language, to deceive.

"The great danger of lying is not that lies are untruths, and thus unreal, but that they become real in other people's minds. They escape the liar's grip like seeds let loose in the wind, sprouting a life of their own in the least expected places,...."

Christine Leunens

Caging Skies, 2004

Chapter Three

Psychological Rudiments

> "We all inevitably present a version of ourselves that is a collection of half-truths and exclusions. The way I saw it, the truth was too complicated, whereas the well-chosen lie would put everyone's mind at ease."
>
> Caroline Kettlewell
>
> Skin Game, A Memoir, 1999

Lying is embedded in the Pleasure Principle.

Psychological Rudiments

Although psychology – the study of mind and behaviour– owes its historical beginning to a number of Freud's predecessors, prominently the German physician Wilhelm Maximilian Wundt (1832-1920) and the American psychologist William James (1842-1910), it was arguably Sigmund Freud, the Austrian neurologist (1856-1939) that put a recognizable face on the word 'psychology' by founding the practice of 'psychoanalysis' – the study (and treatment) of mental and behavioural disorders through creating an interaction between the unconscious and the conscious elements in the mind.

Freud has reportedly shied away from acknowledging Darwin's considerable influence on his own theories, partly in an attempt to put psychology (psychoanalysis to be exact), particularly in its infancy, on a different footing to biology with which Darwin was associated; but also partly, as is the case with all humans, because of the pride of one's own name in relation to one's own achievements.

Freud has, of course, mentioned, in a number of his writings the significance of Darwin's theories to an understanding of ourselves and our place in nature (Bergman, 2010).[(31)]

In a lecture on the unconscious, for example, Freud, ironically, discussing 'major blows at the hands of science [to] the naïve self-love of men' acknowledges that, "The second blow fell when biological research destroyed man's supposedly privileged place in creation and proved his descent from the animal kingdom and his ineradicable animal nature". [32]

In an article titled 'Freud and Darwinism' Jerry Bergman (2010) points out, "In his writings, Freud referred directly to Darwin and his work over 20 times.... Freud was specifically interested in Darwin's work in the area of psychology - for example, in his book 'Expression of Emotions in Man and Animals' Darwin taught the self-preservation theory, an idea that was central to his survival of the fittest concept. The theory developed by Freud and his followers from Darwinism was based on the idea that all behaviour is the result of a few basic animal drives produced by natural selection to facilitate survival. Darwin argued that all animals have an innate self-preservation instinct (i.e.libido) that included both struggle to survive and the drive to reproduce". [33]

According to the Freudian psychoanalysis the human animal, like other animals, survives through the hormonal urge to recreate, instigated by libido - the sexual instinct associated with the Id and the Pleasure Principle.

Seated in our unconscious, the id is a basic component of our personality and plays a major role in shaping/energizing our overall behaviour, including deceptive behaviour from which we draw pleasure (see chapters 5 and 6).

In Freudian psychoanalysis, the id is an actor in our unconscious and one of the three constituents of our personality – the other two being the Ego and the Superego.

The id is called the animal in humans because it is the desiring force in our nature, physiological as well as psychological; it demands pleasure regardless of consequences (see chapters 5 and 6). In spite of its bad reputation the id is vital to sustaining life, anything from suckling milk at birth to seeking sexual gratification, to avoiding danger; in short, all gratification and every pleasure producing thought or action belongs in the domain of this sexual instinct *(note that the word sexual in Freud is used broadly to mean thoughts, feelings and actions that give us pleasure, satisfaction and happiness in contrast to those that cause stress, frustration and conflict – displeasure);* "The id cannot tolerate increases of energy that are experienced as uncomfortable states of tension", write Hall and Lindzey of the University of California (1978), "Consequently, when the tension level of the organism is raised, either as a result of external stimulation or of internally produced excitations, the id functions in such a manner as to discharge the tension immediately and return the organism to a comfortably constant and low energy level. This principle of tension reduction by which the id operates is called the pleasure principle." [34]

According to Hall and Lindzey, in order to avoid pain and obtain pleasure the id uses one of the two processes: reflex action and primary process.

Reflex actions are in-born and automatic like blinking to remove dust from the eye surface or scratching an itchy mosquito bite. The primary process is the psychological reaction to wish-fulfilment and problem solving such as hallucinating (fantasizing) a desired object or fulfilling a wish by dreaming a verbal reality; that is, lying to oneself or to others.

While wish-fulfilment can be one major originator of lying initiated by the pleasure principle, avoidance of, and escaping from, psychological pain demand speedier involvement of that facilitator to prompt lying. Indeed it is Freud's idea that our mental processes (lying for one) are primarily regulated by the pleasure principle. Freud, in his 'Beyond the Pleasure Principle' expounds, "In the psycho-analytical theory of the mind we take it for granted that the course of mental processes is automatically regulated by the 'Pleasure-Principle': that is to say, we believe that any given process originates in an unpleasant state of tension and thereupon determines for itself such path that its ultimate issue coincides with relaxation of this tension, ie. With avoidance of 'pain' or with production of pleasure". (35)

As we grow older we become conscious of behavioural inhibitions imposed by our sense of moral and ethical values to which societies and their cultures greatly contribute. Psychoanalysis calls these inhibitions the Ego, to which the Reality Principle belongs.

The Reality Principle has enforceable powers and, amongst

other things, has the job of instilling sense into the id and when necessary promising restoration of any lost pleasure, whether it is a car-breakdown on the road or loss of a loved one. The reality principle, "...is the taking into account of external reality in addition to internal needs and urges", say Carver & Scheier of Miami and Carnegie Mellon Universities (2004). [36]

There is yet another safeguarding agent with regards to the id's wild expectations or, for that matter, any miscalculation or erroneous judgement on the part of the Ego itself, and that is called the Super-ego which is also given the unenviable nickname of the 'grandfather', as if it is someone who sits in his rocking-chair somewhere in some corner of our mind ready and willing to discharge his wisdom, warning us against temptations that may distract us from good behaviour.

The Pleasure Principle, therefore, is a life preserving mechanism, the true motivational force in our psyche which, itself, goes through five psychosexual stages of personality development (bearing in mind, as previously mentioned, that the word sexual in Freud is used broadly to mean thoughts, feelings and actions that give us pleasure, satisfaction and happiness in contrast to those that cause stress, frustration and conflict – displeasure).

At birth and during infancy our behaviour is controlled solely by the id and its pleasure seeking principle: The oral stage which is when the id is centred around the mouth and directs us to the mother's breast for a feed. This libido driven instinct gradually moves away from the mouth at age one or thereabouts to the anus from where the child derives the pleasure of defecating (also the presumed onset of the ability to lie). This is the stage where the Ego makes its first appearance and gives the child a sense of identity away from the mother-dependency, and in conflict with parental authority.

From three to perhaps six years of age is what Freud calls the Phallic stage, the times when the child becomes aware of the difference in sex anatomy. This stage is also marked by the emergence of conflicting feelings and sentiments in the child's life – jealousy, fear, competition and rivalry. Arguably, it is also at this stage of growth that the tendency to lying makes its fluid appearances.

Freud calls the fourth stage of development, anything from phallic to puberty, the Latency stage when the libido tendencies give way to more responsible matters of life, acquiring skills and knowledge. During this stage libido takes a back seat temporarily.

Finally, the Genital stage that runs from puberty onwards, covering sexual experimentation, relationships: personal, social, cultural, occupational and economic. From this stage onwards humans also expand on, and fine tune, their deceptive skills to stay one step ahead of anything that challenges the status, immunity and the progress of their relationships and personal well-being.

It should be noted, though, that while the Freudian concepts of psychosexual development sound explanatory enough, there are doubts amongst experts about their scientific accuracy.

In summary, according to Freud, life, ie. psychosomatic life, pertaining to the relationship between mind and body, consists of two constituents: Pleasure and displeasure.

As we progress in age, the control mechanisms emanating from our conscious mind, namely the ego and the superego (our awareness of the inhibitive/interfering forces such as civil and legal laws, sociocultural considerations, etc.) which are, in turn, the agents of the reality principle, become more influential in our personality viz-a-viz the pleasure principle.

The reality principle is there to put us in touch with the external world and normalize our relationship with it, not the least our socio/ethical relations in all aspects of life. However, the reality principle remains in constant tug-o-war with the pleasure principle which sees itself as a more important agent when it comes to motivating living, satisfying one's physical and psychological needs, eg. loving oneself (narcissism), enjoying food, enjoying touch, smell, company, beauty, scenery, speed, vitality, winning, fantasies, stories, success and everything else that makes life worth living (pleasurable) for us.

The reality principle, on the other hand, views its role as importantly, arguing that it must uphold the rules of conscience, awareness, as without it existing, extreme behaviour desired by the pleasure seeking activities of the id can endanger life. But the pleasure principle keeps using all possible tricks in its armoury to bring to us pleasurable gains; and deception is no exception; "Charles Darwin suggested that children as young as thirty months are capable of lying after seeing his young son trying to deceive him", writes Romeo Vitelli. (37).

Of course deception is a generic terminology and covers a wide range of trickery of which lying is one, albeit one that falls in the verbal , that is, psycho-lingual, category; 'When Does Lying Begin?' asks Romeo Vitelli in the article bearing that title, with the sub-heading 'How early do we learn to lie? And what purpose does it serve in young children?' (38) "It's tempting to argue", explains Vitelli, "that lying is also linked to creativity since the ability to create fiction often relies on the same cognitive capacity, the ability to balance more than one reality in their head (which is what liars do when they create a fictional version of events to match with the truth) becomes easier. It also means being able to recognize the difference between fiction and reality, much like what they watch on television as well as being able to create new stories." (39)

The cognitive capacity that Vitteli mentions comprises, in short, of learning and thinking which arguably start at birth primitively through senses, "When only a few days old" explains Thomas Berndt (1992), referring to an earlier research by

Jennifer M Cernoch & Richard H. Porter, (1985), "breast-feeding infants will turn toward a nursing pad their mother had used rather than toward another mother's pad. Infants could behave this way only if they recognized - that is, remembered - the smell of their mother's nursing pad". Berndt goes on to say that newborns respond, not only to smell but also to other sensory stimuli: visual recognition 'to buzzers or other sounds, and to the touch of various objects'. [40]

Interestingly, Berndt postulates that, "Even a fetus in the womb will show habituation [loses interest in a repeated stimulus]. If a vibrating object is placed on a mother's abdomen when her fetus is about seven months old, the fetus will start to jump around. If the vibrator is removed for a short time and then placed again on the mother's abdomen, the fetus will again move. But if this sequence is repeated several times, the fetus will gradually quit reacting to the vibrator (showing displeasure). That is, the fetus will habituate to the stimulus. This habituation [displeasure] shows that the fetus remembers something about previous experiences with the vibrator". [41]

One person whose name is closely associated with, and recognized as, the founder of the theory of cognitive development in children is the Swiss psychologist and genetic epistemologist, Jean Piaget (1896-1980).

Piaget, like Darwin before him, thought of adaptation as the key to intellectual development of organisms; "The basic tenet of Piaget's theory of development", says Bruce Tuckman of Florida State University, "is that organism develops schemata that enable it to continue to function in that environment. The very essence of life is a continuing and repeatable interaction between the organism and its environment that enables the organism to function". [42]

Amongst a plethora of Piaget's discoveries was the stage of preoperational thought between the ages of two to seven. One phase of this stage is Egocentrism when the child is inclined to think that people around her/him think the same as she/he (in Hide and Seek, for example). Any deviation from this outlook is considered false in which case the child uses ploy (lies) to motivate adoption of her/his version of reality. Associated with egocentrism is Piagetian notion of Animism.

The Latin derived word, Animism, commonly signifies the religious view that all existence: rocks, plants, animals, wind, water, man-made objects such as painting, statues, even words are alive. For Piaget Animism "meant the tendency to assume that non-living (inanimate) objects such as the sun or the wind have properties of living things. In particular, these objects have motives, feelings and intentions that affect their behaviour". [43]

At the animistic stage of cognitive development "young children confuse the physical world with the purely psychological world".[44]

In order to establish the theory of animism, Piaget and his research team developed questions (1929), and asked them to a group of eight year old boys and girls. For example, they asked whether the sun was alive. 'Yes' the children responded 'because it makes daytime', or whether clouds were alive; 'Yes, because they make it rain' etc. [45]

By way of fun my wife and I enacted the Piagetian animism on her two and a half year old nephew.

On his birthday my wife put a chocolate birthday cake in a room fitted with a small CCTV camera (as well as the recording mobile phone) telling Newton that she was going to get the candles/cutlery etc., and that he was not to touch the cake until we all returned to the room. We then shut the door and left to watch Newton's behaviour on the monitor in the living-room. It was quite entertaining to watch Newton's behaviour from his anxious gazing at the shut door while smelling the cake, to finger scraping the cake from the bottom edges of the silver laminated cardboard base and moving the shattered pieces, palmful, carelessly to his mouth, to bending down to smooth down the by-now jagged edges of the cake by a licking action that was smearing chocolate all around his mouth and on his nose.

'Good boy, Newton! You didn't touch the cake, did you?'

Newton nodded 'No' in slow motion as he concentrated on

wiping his chocolate covered fingers on his new white pants.

'Did Apple (our Maltese dog) come inside?'

Newton reassuringly gave brief repetitive jerks of a 'yes' nod, oblivious of the fact that the window and the door to the balcony were shut and the dog could not have come into the room.

'Did Apple eat this part (of the cake)?'

Another confident Yes nodding; he then turned his head away to avoid eye contact.

'Did Apple rub the cake around your mouth?'

'Did the cake tell Apple no one was allowed to touch it?'

A short moment of thoughtful contemplation was followed by repeated hesitant (yes) nods, then a firm (No) shake of the head. Pointing a finger towards the window then quickly raising it to his mouth and resting it on his lower lip, eyes wide open, he exclaimed, as if accusingly: APPLE! (The dog's name).

In some cultures, in fact in most cultures, animism is rooted in the belief systems of communities (and in the individuals belonging to those communities) in which case it is no longer a childhood fallacy, rather a life-long cultural credence. For example, a group of Aboriginal elders gather around an eight hundred year old tree in the Kaurna country in Adelaide (the capital city of the State of South Australia, in Australia) every Sunday or so to comfort the tree in its old age, talk to it, share stories with it, hug it and wish it good health.

Religious animism is everywhere for us to see whether it is a silver plated (Christian) cross around one's neck, or the gold plated metal grills around the shrine of some miracle giving saint to which pilgrims tie themselves for days to get the deceased saint attend to their needs – financial, emotional and health needs.

Linked to the phenomenon of animism is the concept of Artificialism or symbolic naturalism; the view that realism is essentially artificial, only symbolically true. For example, the toy fire truck is as real to the child as the real one, and if it doesn't make the sounds that the real one makes the child adds his own sound to it to make the pretend object feel real.

In the article posted in 2018 under the heading 'The Preoperational Stage of Cognitive Development', Saul McLeod informs us of the key features of this stage, namely the symbolic representation: "The early preoperational period (ages2-3) is marked by a dramatic increase in children's use of the symbolic function. This is the ability to make one thing – a word or an object – stand for something other than itself. Language is perhaps the most obvious form of symbolism that young children display....Toddlers often pretend to be people they are not [e.g. superhero, policeman], and may play these roles with props that symbolize real life objects. Children may also invent an imaginary playmate…. During the end of this stage children can mentally represent events and objects [the semiotic function], and engage in symbolic play.".[46]

Children's awareness of the effectiveness of verbal and body language such as shedding crocodile tears, screaming hysterically, throwing tantrums, sitting put and various other blackmailing exercises are testimony to the universality of natural deceptive behaviour amongst children (and in some cases in adults as well). Verbal language in particular is an invaluable tool in the practice of lying regardless of age, race, culture, nationality or gender.

Hunter in the dark.

"I wrote home to say how lovely everything was, and I used flourishing words and phrases, as if I were living in a greeting card – the kind that has a satin ribbon on it, and quilted hearts and roses, and is expected to be so precious to the person receiving it that the manufacturer has placed a leaf of plastic on the front to protect it."

Jamaica Kincaid
Lucy, 1990

Chapter Four

Linguistic Rudiments

I'm sorry
I love you
You're beaut ful
It's my fault
I forgot
I don't know
Whats for supper?
Who_me?
I'm not lost
BURP!
Duh!

'Language ushered in a new phase in timeless struggle between the forces of deception and detection. Loading the dice heavily in favor of the former, it enabled human beings to misrepresent reality much more effectiverly than had previously been possible.'

David Livingstone Smith
WHY WE LIE, P.110

Lying is a semiotic phenomenon. As a word stands for the thing that it is not, a lie stands for the truth that it is not.

Linguistic Rudiments

Animals use deceptive communication to survive - for reproductive purposes, for predatory (food) or anti-predatory (distracting predators) reasons or simply for the purpose of territoriality. That said, as is the case with humans, not all animal communication is intended for the purposes of deceiving. Animals use signalling to communicate information: the dancing of bees to indicate the direction to the hive or the distance to the nectar, for instance; the elaborate whistling and grunting of whales and dolphins, or the bird call-signalling to fly away, or the dog looking expectantly to play fetch or scratching the door to come inside. [(47)]

Such acts of communication, universal as they are, happen at primary level. Human language, an aspect of their communication repertoire, is a complex evolutionary mechanism.

Humans too use deceptive communication to survive obstacles, only in far more complex ways than other animals and for far more complex reasons. The reproductive purposes still apply in our manners of courtship. Predation in humans may not necessarily be for food as such (although that could be too), but, semiotically speaking, objects of our desire.

Our anti-predatory actions are the strategies that we develop to fend off competition; our territoriality lies in all the efforts we make to prevent intrusions and interruptions. This is where most, but not all, similarities between human and animal deceptive tendencies end.

Our language is at the service of our relationships, which are innumerable: our relationship with ourselves (thoughts, feelings, beliefs), with others, with nature.

Maneuvering our way through the maze of these relationships requires calculated (lateral) ploys sometimes spontaneously, sometimes not.

It would be an impossible task to track the evolutionary history of a particular linguistic phenomenon, lying for instance, within the repertoire of human communication. Research suggests that the tendency to deceive, in the biological world at least, predates the emergence of human language. (48)

Darwin, to whose theories on the biological foundation of deception we have been referring in the previous chapters, did not engage seriously in an attempt to explore the origin of human language until the publication of his book, 'The Descent of Man' (1871). There are, however, indications that as early as 1830's Darwin was expressing scanty thoughts on the subject of communication in animals and humans (Barrett et al ed.1987). (49)

Darwin viewed language as a dual process of concept for

mation and articulation.

Concept formation, according to Darwin, is not uniquely a human ability; animals too have the ability to link concepts to sounds, gestures and facial expressions to communicate messages. What is uniquely human though is the 'Articulate Language'; language that works on syntactic and semantic formations whether in children's rudimentary utterances or the elaborate scholarly speech (the written language is a different matter) - a process in human cognitive evolution.

As Philip Kerr says in his introduction to the book, 'The Penguin Book of Lies' (1990), "It seems unlikely that man learned to speak before he learned to lie, but rather that he learned to lie at the same time as he learned to speak. Thus the capacity for lying is essential to the speech function. Just try imagining the man who had no capacity to lie: one might just as well say that a man was dumb and could not speak at all".[50]

The Movie, 'The Invention of Lying' (2009) directed by Ricky Gervais and Mathew Robinson, a fantasy romantic comedy, is a good representation of what the world would look like if lying did not exist – dysfunctional.

Since language and thinking are tightly interwoven (Kaplan D. & Manners R. 1972),[51] we may as well accept that the history of lying is as old as the history of thinking – that is, the history of Homo Erectus when humans, by some speculative accounts, used sophisticated noises to bond together in a group and to discuss plans of travel or of attacks – collective bird or animal migration does not happen by accident. (52)If so, then the theory that human language developed primarily for social bonding holds sway; so does the assumption that deceptive language developed as a linguistic function for the purposes of group cooperation to win over adversarial conditions.

Conrad P. Kottak, in his book 'Anthropology: The Exploration of Human Diversity' (1974) adds a caveat to his own assumptions, saying, "I also speculated that by the Homo Erectus stage of human evolution some communication more advanced than a call-system was one of the adaptive advantages associated with cooperation and learning". (53)

Kottak further adds, "You must understand, however, that this was speculation. There is no way of ascertaining whether the communication systems of Homo-Erectus would qualify as language. In fact, all statements about the ultimate origin of language are speculative". (54)

The debate over the origins of language seems constant, but some researchers (Scott-Phillips, 2010) have attempted to abate the debate by suggesting that the conversation should be directed mainly towards 'Communication and Evolutionary Functionality' of language.

Ahough he too gets entangled in the pros and cons of the argument over the theory of 'Natural Selection' and the origin/s of language, Scott-Phillips, citing Ruth Garrett Millikian(1984), agrees with her distinguishing between Direct and Derived functionality in relation to the deceptive use of language and its generational regeneration.

Scott-Phillips claims that, "This distinction allows us to make sense of the various claims [on the evolutionary origin of language]: communication is the direct function of language, while acts such as gossip, courtship, hunting and so on are various derived functions of language. Put another way, the ability of language to influence the behaviour of others explains its reproduction from one generation to the next, and it achieves this through the production of linguistic behaviours that we variously call gossip, chat-up lines and so on". (55)

In an informative article on the topic of 'Deception as a Derived Function of Language' Nathan Oesch, (2016), expounds, "Language may be one of the most important attributes which separates humans from other animal species. It has been suggested by some commentators that the primary biological function of human language is to deceive and selfishly manipulate social competitors". (56)

Oesch's article navigates through sources on human social behaviour, comparative animal behaviour and developmental psychology to suggest that "deception shows clear signs of a derived function for language. Furthermore...

across most human and non-human animal contexts, deception appears to be utilized just as often for prosocial and social bonding functions, as it is for antisocial purposes". [57]

Further, Oesch, with reference to other research, distinguishes between the tactical and the functional acts of deception; the former being predominantly a prerogative of animal behaviour and the latter that of humans.

Oesch mentions documented cases of tactical deception such as alarm calls and warning cries by animals in order to increase their own chances of food supply or reproductive opportunities. As an example, Oesch mentions the case of the male antelopes that alarm-snort deceptively to prevent the females within their territory from straying too far away from the group, and by doing so risk the chances of becoming either attracted to, or attracted by, the males from other groups.

Or the case of Capuchin monkeys that sound alarm-calls normally reserved for predator sighting in order to attract dominant members of the group away from the food source and by taking advantage of their absence steal food and run away with it.

Such self-serving functions in humans, through prosocial or antisocial lies, are carried out by means of verbal language (not exclusively of course) - tales of Pinocchio are very famil iar examples although historical anecdotes/ records attributed to Niccolo Machiavelli (1469-1527) come a close second to our own daily experiences with a lot of people, not the least with politicians, as well as other leaders, who take advantage of language, regularly and masterly, to convert the truth into believable untruth.

As an example of our daily experiences, Oesch cites research by Toma et al., (2008) conducted to study deceptive claims by Americans on internet dating sites whereby 90% of males and 75% of females did not tell the truth at least on one physical characteristic (for example, claimed greater height or less weight). (58)

Karl R. Popper, the Austrian/British philosopher (1902-1994) takes lying as a derived function of language even further in a book edited by Philip Kerr, titled ' The Penguin Book of Lies'.

In the introduction to the book, Kerr, with reference to Popper and the famous American linguist and philosopher, Noam Chomsky (born 1928), writes: "Popper's view of the origin of liars would seem to be strongly supported by Chomsky's belief that we are genetically pre-programmed to master the use of language. Chomsky holds that language grows in an infant's mind just as it has grown arms and legs while still in its mother's womb. Inherent in that growth of the language function must be the capacity for lying. Indeed, says Popper, lying 'has made the human language what it is: an instrument which can be used for misreporting almost as well as for reporting". [59]

We may put Chomsky's views in perspective by returning to Darwin. Darwin too recognized that humans are hard-wired to the ability to shape their instinctive sounds (born with them) into acquired communicable formulations but he did make a distinction between pre and post babbling stages of language development.

At pre-babbling stage humans share their instinctive sound making ability with animals: "The sounds uttered by birds offer in several respects the nearest analogy to [human] language, for all the members of the same species utter the same instinctive cries expressive of their emotions; and all the kinds that have the power of singing exert this power instinctively; but the actual song, and even the callnotes, are learnt from their parents or foster parents. These sounds...are no more innate than language in the man. The first attempt to sing may be compared to the imperfect endeavour in a child to babble".[60]

The word 'communicable' is used here to suggest that humans, from birth, begin to communicate (send and receive) meanings by making sounds - what we generally call 'cry', to which parents, mothers in particular, respond by performing some action eg. back tapping, breast feeding, cradling, murmuring/ talking to the baby, kissing, etc.,. This phase although communicable is regarded by Darwin as 'inarticulate'.

We may also speculate with some degree of certainty that since exchange of meaning is involved in this phase, the baby would have the ability to use deceptive cries (crocodile tears, for instance) to have their feelings or needs attended to. This, so called the vowel phase, begins to develop into the consonant phase (babbling) when the baby reaches four to six months of age, generally speaking, with the enunciation of P and B sounds and (the consonant phase) progresses as the baby's vocal organs strengthen through their use .

One other noteworthy remark can be made in respect of the baby's deceptive intent at, during and after the inarticulation or the cry phase; and it relates to Freud's view that the baby is born with the innate sex drive, libido, whereby the baby bonds with the mother (an act of courtship), feels secure and belonged - we know that babies, like some animals, distinguish the smell of their milking mother from other females with whom they come in contact; hence the baby's head turning, or kicking and other acts of rejection of strangers to show displeasure, a behaviour that we commonly associate with shyness.

Any event that comes close to putting this bondage and security at risk produces a sense of fear combined with jealousy that the baby articulates through crying, mostly deceptive crying (crying without tears) or the so called crocodile tears. As soon as the threat is warded off and security restored the crying ceases instantly. We commonly experience this reaction when the baby is picked up, or approached to be picked up, or even reached out to for hugging, by a stranger.

Fear and jealousy, as two innate mechanisms of self-preservation and self-actualization, grow in us, and alongside language, through the babbling age of 7-8 months, through the more intelligent and intelligible age of talking (usually at age two) and beyond - throughout life.

Self-preservation and self-actualization are motivational enough for us all to use whatever linguistic devices, verbal and/or non-verbal available to us, deceptive for one and a major one at that, when it comes to matters of strong emotions such as in relationships and matters of survival – consider these in the broad contexts of private and public life from romance, to friendship, to finance, education, promotion and all that living involves. Take the simple case of the child who was asked by her mother to remain at home with another sibling while the mother was preparing to go shopping with her older daughter. The child's use of the primitive tactic of crying resulted in her being taken along – a ploy that resulted in self-actualization. In short, self-preservation and self-actualization are two generic causes of linguistic deceptive behaviour in humans. Lying and language are contemporaneous.

Dancing lies of love.

"When it comes to controlling human beings there is no better instrument than lies. Because, you see, humans live by beliefs. And beliefs can be manipulated. The power to manipulate beliefs is the only thing that counts."

Michael Ende

The Neverending Story, 1979

Chapter Five

Mythical and Philosophical Rudiments

"The great enemy of truth is very often not lie – deliberate, contrived and dishonest – but the myth – persistent, persuasive and unrealistic. Too often we hold fast to the clichés of our forebears. We subject all facts to a prefabricated set of interpretations. We enjoy the comfort of opinion without the discomfort of thought."

John F. Kennedy
Speech in Yale University, 1962

'Humans are pattern seeking, storytelling animals, in search of deep meaning behind the seemingly random events of day-to-day life.'

Michael Shermer

Why People Believe Weird Things/1997

Mythical & Philosophical Rudiments

Film Is Real. Film Is Not Real.

I am sitting in a chair in cinema 3 at the Hoyts, in the middle of the middle row, half way up the salon watching a movie. The world has become the space of the chair in which I am sitting. The space is the universe of the poetic reality on that screen in front of me – nothing else exists around me, above me, beyond the film about me; without me viewing it, the film will lose me and itself – its animated reality. I am the film; my fantasy eyes are touching the screen and finding that it is real. My emotions, views and values are synchronized with the screen, Wifi connected to it, amplified in this intriguing space. I am experiencing the mutation of an it to an *it* – a reality to a *reality*: **the film is real because it is a film; the film is not real because it is a film** – the object of its own subjectivity. Nonetheless, I am sitting in a seat in the middle of the middle row, halfway up the movie theatre caught in the dialectics between le éxistence de imaginaire et le éxistence de réal.

Allan Nanva

Mythologies of Late

Myths are false truths - false because they are fabrications; truth because they have formed and continue to form beliefs and practices of a great many people, arguably the whole humanity, in one way or another: superstitions, astrology, exorcism, sorcery, vodun, cosmology, medicinal rudiments (eg., administration of placebo), witchcraft, mediumship (or Ouija, the practice of mediating communication between the spirits of the dead and living human beings).

The question whether myths, tales, fables and pious fabrications, particularly in religious texts, should be taken literally or figuratively is too vast a topic for one chapter. What can be accommodated in this space is putting in context certain historical views that shape the belief systems and cultural practices of people everywhere every day.

That said, I feel there may be occasions when I may have to resort to science, again, to help me demonstrate the rudimentary biological roots of (mythical) beliefs. After all, we are the beneficiaries of modern science which busts myths and explains origins of phenomena, helping us to unravel mysteries of life (and death).

However, modern science is only 200 or so years old whereas the history of human race as we know it, based on the anatomical fossils found in Ethiopia, goes back some 200,000 years. On that account humans have had 180,000 years or so to think of mythic explanations for matters life and death - those ancient in-brained beliefs that we have inherited and we are finding it impossible to exchange with science especially when myth merchants in various fields continue to nurture their continuation, even snow-ball their original forms as well as their impacts.

I shall make 'suspension of disbelief' my point of departure: what is in myths (fairy tales, movies, novels, religious stories in holy books, magic shows) that makes them inherently believable even when something in the outdoors of our brain tells us they are not true?

The phrase suspension of disbelief prevalently used in film and literary analyses as well as in psychiatric and related medical texts was coined by the British Romantic poet Samuel Taylor Coleridge (1772-1834), supposedly inspired by the Roman philosopher and orator Marcus Tullius Cicero's (106BC-43BC) phrase 'Assensus susepensione' to suggest that humans are vulnerable to taking the fantastic for the real if composers of works of art (poets in his case) infuse 'semblance of reality' into tales of human interest.

In chapter xiv of his 'Biographia Literaria' (1817) Coleridge explains that by the phrase 'willing suspension of disbelief for the moment', specifically talking about notions of Poetic Faith and his two poems: 'The Rime of the Ancient Mariner' (1797-98) and 'This Lime-tree Bower My Prison' (1797), he intends to invoke in the responder the pleasure of believing that the untrue is true for the period that they are engaged with the (fabricated) text.

The story behind Coleridge's poem, 'This Lime-tree Bower My Prison' and the poem itself are a good demonstration of his ideas on the 'willing suspension of disbelief'.

The poem is an imaginative picture (imagine a movie), painted emotively, of a situation Coleridge experienced in his village of Somerset while he was visited by a group of his friends including the famous British poet William Wordsworth.

It was a long-awaited visit and the party had planned to go for a long walk around the natural scenery of Somerset. However, on the morning of the planned walk in the early hours of that day. Coleridge injured his leg badly in his garden and was not able to accompany the group in the walk.

While sitting in this garden bower, friends gone, Coleridge imagines his friends' whereabouts and their experiences, metaphysically, vicariously joining the group as if present and walking with them over hilltops, looking at the inspiring views of surrounding countryside, down to the ocean where they see a ship, appreciating its majestic beauty; then they pass a magnificent church, etc.

> Well, they are gone, and here must I remain, … They, meanwhile,
>
> Friends, …
>
> On springy heath, along the hill-top edge,
>
> Wander in gladness, and wind down, perchance,
>
> To that still roaring dell, of which I told;
>
> The roaring dell, o'erwooded, narrow, deep,
>
> And only speckled by the mid-day sun;
>
> Where its slim trunk the ash from rock to rock
>
> Flings arching like a bridge; that branchless ash,
>
> Unsunn'd and damp, whose few poor yellow leaves
>
> Ne'er tremble in the gale, yet tremble still, Fann'd by

the water-fall!

and there my friends

Behold the dark green file of long lank weeds,

That all at once (a most fantastic sight!) Still nod and drip beneath the dripping edge

Of the blue clay-stone.

Now, my friends emerge

Beneath the wide wide Heaven—and view again

The many-steepled tract magnificent

Of hilly fields and meadows, and the sea,

With some fair bark, perhaps, whose sails light up

The ship of smooth clear blue betwixt two Isles

Of purple shadow! Yes! they wander on

In gladness all; ...

A delight

Comes sudden on my heart, and I am glad

As I myself were there! Nor in this bower....[61]

When our brain receives stimuli from our five senses (Aristotle's classification), it forms a pattern of meaning compatible with our previously coded belief patterns: environmental, ideological or cultural. Our belief patterns are in constant interaction with our emotions which are also tied to our memory. Such interactions arise from neurological and biochemical sources.

The study of the evolutionary relationship between the brain and the mind is fascinating as, to some extent, a review of such topics in previous chapters would have shown.

That said, understanding how immaterial entities such as thoughts, memories, dreams, beliefs and language arise from a biological matter called the brain demands a brief review. We may explain this process through the example of watching a movie – a fabricated reality:

Let us imagine ourselves watching a film in the movie theatre because of its designed and controlled atmosphere: space design, lighting and acoustics. Let's, for the purposes of this example, take one emotionally charged scene in this dark environment and follow its psychosomatic trail of actions and reactions (in the brain) at a speed inexplicable.

Upon viewing the scene, the stimuli (audio/visual) activate the brain's limbic system – a group of interconnected structures namely the hippocampus, amygdala, the fornix, cingulate gyrus and others. There, in the limbic system, the audio/visual information goes through a process of selection and combination, gets interconnected with other spontaneously received and/or memory stored information catalysing release of a range of bio-chemicals known as ligands – comprising of peptides, hormones, etc.

By means of certain to-and-fro electrical signals (dendrites that bring the information to the cell and axons that take information away) the ligands reach billions of targeted neurons (information carrying nerve cells) in multiple interconnected areas of the brain where they become impressions, patterns of meaning that form our (kaleidoscopic) perceptions: linguistic, affective (emotional) or ideological. The repetition of similarly (sense) stimulated brain activity, through various cultural practices, for instance, reinforces our (emotionally held) views the intensity of which can be maximized, minimised or even replaced via neutralizing mechanisms such as administration of (medicinal/psychotic) drugs or psychological (mind altering) programs such as hypnosis, mindfulness, or a combination of such programs exemplified in the medical practice of treating patients by means of the placebo effect - Placebo is fake medicine/healing practice that produces real therapeutic results because the patient is led to believe its medicinal or healing effects. The practice mimics Aristotle's principle of imitation of reality, the Aristotelian mimesis, (62) (meaning imitation, a model of truth, mimicry) particularly his notion of catharsis.

Aristotle, discussing the constituent elements of tragedy in his book 'Poetics', defines catharsis as an act that, "By evoking pity and terror it brings about the purgation of those emotions". (63) In Aristotle, as well as in psychoanalysis, catharsis is purification of the mind by recalling suppressed emotions via imitated reality.

Presented as models of truth: advertising, online propaganda, political speeches, religious sermons and rituals, so called education camps, ideological debriefing programs and numerous other processes take advantage of human susceptibility to deceptive manipulation. [64]

Suspension of disbelief is an act of self-deception; it is our pleasurable vicarious participation in imaginary constructs - Freud's pleasure principle in action; his reality principle in retreat.

When we are viewing a movie or reading a novel (or the stories read to us when we are two years of age or even younger) our conscious and unconscious faculties lose demarcation, each yielding its resources to the other in an act of total cooperation, thus allowing our mind to re-write the code of our reality, a process that begins around age two (or even earlier; in fact, some may argue that we get hard-wired to it in the womb) and continues throughout our lives - imagine the baby that from the moment of birth takes the dummy for the mother's breast-nipple, the child that takes the toy fire-truck for real, the adult who fantasizes a sexsual intercourse.

Moments of our lives are filled with stimuli of all sorts: verbal, sense triggered, etc., which swing us between reality and unreality, force us into and out of dreams and fantasies - when we are in the lecture room, engaged in conversation with others, watching television, driving the car, riding the bike or the horse, lying down on the sofa or cooking, or walking in the street passing the perfume store or the fruit shop or the restaurant, passing that lamp post where we met her/him, listening to the religious preacher or reading the holy book's scripture that fly us via Amygdala to hell or to heaven, the toys that we animate in childhood, stories that are read to us or we read in young age, the Santa Claus whose presents down the chimney under the Christmas tree we await, the games that we win without playing them, the sexual relationships that we fantasize, the space travel that we journey, the castle that we occupy, the good looking man or woman in the supermarket that we fancy, the million or so dollars that we don't have but in the bank account we keep, that nasty muscular aggressive thief that we catch in our house whose hands and legs we tie up in our dream while we are asleep, the mountains over which we glide in the air sleeping: all our day dreams and night dreams and more. Whoever said life is but a dream.

The history of suspension of disbelief - willingly or otherwise accepting lies as facts, goes back to distant past when primitive humans sought supernatural explanations for phenomena: physiological (eg. illnesses), seismological and environmental. Over aeons myths evolved into beliefs some of which, religious beliefs for instance, are so ingrained in our ancient brain that are now arguably as much physiological (brain matter) as they are psychological (mind matter).

In Greek mythology Dolos (the fraudster) is the son of Gaia (earth) and Aether (sky). He learns his fraudulent skills under the mastership of Prometheus (who tricks gods and makes man from clay), befriends Pseudologoi (spirits of lies and falsehood) and in partnership with Apate (goddess of deceit) brainwashes gods into telling lies. Pseudologoi were children of Eris (hostility) and had brothers all of whom had names associated with consequences that lying could bring about: Limos (starvation), Algea (pain), Ate (destruction), Phonoi (Murders), Hysminai (war), Dysnomia (chaos).

In his 'Republic' (380 BC), Plato engages Socrates in conversation with a number of debating partners: Thrasymachus, Cephalus, Polemarchus and Glaucon, to discuss a range of topics related to the utopian society that they are envisaging. One such topic is the role of lying in the society (ie.the Republic).

The topic of lying, more broadly the issue of appearance versus reality, also noticeable in pre-socratic records in Greek history, had been debated by Xenophanes (570-475 BC), Empedocles (495-436BC) and Parmenides (515-440 BC) long before 'Republic' (although Parmenides appears as a major contributor to debates in Plato's 'Dialogues' written between 399-387 AD).

On the matter of natural human vulnerability to deception - taking appearances for reality - for instance, Plato in book II of 'Republic' tries to prove this point through the parable of a dog-

Socrates addressing Glaucon, Plato's older brother:

S. Here then I was perplexed, but having reconsidered our conversation, I said, we deserve, my friend, to be puzzled, for we have deserted the illustration which we set before us.

G. How so?

S. It never struck us that after all there are natures, though we fancied there were none, which combine these opposite qualities.

G. Pray where is such a combination to be found?

S. You may see it in several animals, but particularly in the one.... For I suppose you know that it is the natural disposition of well-bred dogs to be perfectly gentle to their friends and acquaintance, but the reverse to strangers.

G. Certainly I do.

S. Whenever they see a stranger, they are irritated before they have been provoked by any ill-usage; but when they see an acquaintance they welcome him, though they may never have experienced any kindness at his hands. Has this never excited your wonder?

G. I never paid any attention to it hitherto; but no doubt they do behave so.

S. Well, but this instinct is a very clever thing in the dog, and a genuine philosophic symptom.

G. How so, pray?

S. Why, because the only mark by which he distinguishes between the appearance of a friend and that of an enemy is that he knows the former and is ignorant of the latter. How, I ask, can the creature be other than fond of learning when it makes knowledge and ignorance the criteria of the familiar and the strange? [65]

The question of the so called Noble Lie (the word noble has been refuted as a mistranslation of the Greek word) and the Lie in Soul - soul in Plato is the individual; and the individual is the embodiment of Reason - searching truth, Spirit (desiring honour) and Appetite (cravings for physical pleasure and material possessions), appears in different sections of 'Republic'. In book II Plato delves into whether what we believe is true and if so how can we be certain; how can we be sure that our beliefs do not originate from our cultural (including religious) and environmental influences and upbringing.

In 'Republic' Plato distinguishes between two types of untrue belief: one that the person is aware of what she/he is expressing/presenting is untrue; this Plato gives the generic term story-telling (think of fiction, poetry or in the context of modern times, movies).

The other type is when the person holding the belief is unaware of it being false. This person is holding a lie in the soul; she/he is genuine in the act of believing but ignorant of it not being factual. This type of lie is dangerous in the sense that it gets passed on from generation to generation through cultural practices (myths as faith) and upbringing, ingraining itself in one's unconscious.

While in conversation with Adeimantus (Plato's brother) Socrates disputes Protagoras' view that man, made in the image of God and having His soul, is all knowing and therefore when he (man) believes something to be true it must be so:

> **S.** Would a god consent to lie, think you, either in word or by an act, such as that of putting a phantom before our eyes?
>
> **A.** I am not sure.
>
> S. Are you not sure that a genuine lie, if I may be allowed the expression, is hated by all gods and by all men?
>
> A. I do not know what you mean.
>
> **S.** I mean that to lie with the highest part of himself, and concerning the highest subjects, is what no one voluntarily consents to do; on the contrary, every one fears above all things to harbour a lie in that quarter.
>
> A. I do not even yet understand you.
>
> **S.** Because you think I have some mysterious meaning, whereas what I mean is simply this: that to lie, or be the victim of a lie, and to be without knowledge, in the mind and concerning absolute realities, and in that quarter to harbour and

possess the lie, is the last thing any man would consent to; for all men hold in especial abhorrence an untruth in a place like that.

A. Yes, in most especial abhorrence.

S. Well, but, as I was saying just now, this is what might most correctly be called a genuine lie, namely ignorance residing in the mind of the deluded person. For the spoken lie is a kind of imitation and embodiment of the anterior mental affection, and not a pure, unalloyed falsity; or am I wrong?

A. No, you are perfectly right.

S. Then a real lie is hated not only by gods, but likewise by men.

A. So I think.

S. Once more: when and to whom is the verbal falsehood useful, and therefore undeserving of hatred? Is it not when we are dealing with an enemy? Or when those that are called our friends attempt to do something mischievous in a fit of lunacy or madness of any kind, is it not then that a lie is useful, like medicine, to turn them from their purpose? And in the legendary tales of which we were talking just now, is it not our ignorance of the true history of ancient times which renders falsehood useful to us, as the closest attainable copy of the truth?

A. yes, that is exactly the case.

S. Then on which of these grounds is lying useful to god? Will he lie for the sake of approximation, because he knows not the things of old?

A. No; that would be indeed ridiculous. [66]

While Socrates legitimatizes fiction and poetry as acceptable approximation of truth, he warns us not to take such crafts and their craftsmen as believable-

"Whenever a person tells us that he has fallen in with a man who is acquainted with all the crafts, and who sums up in his own person all the knowledge possessed by other people singly, to a degree of accuracy which no one can surpass – we must reply to our informant that he is a silly fellow, and has apparently fallen in with a juggler and mimic, whom he has been deceived into thinking omniscient because he was himself incapable of discriminating between science, and ignorance, and imitation".[67]

Plato in 'Republic', and his student Aristotle (384BC-322 BC) in 'Posterior Analytics' (350BC), tried to explain the difference between belief and true knowledge. However, an element of indeterminacy exists in both texts in the context of this distinction although one may walk away from the two texts with the common-sense understanding that belief is opinion (doxa) - supposition (hupolepsis) to be exact- resulting from lack of true knowledge. As a consequence we believe what we know; in other words (a) belief and knowledge become interchangeable, and (b) belief already has a component of suspension of disbelief in it in the sense that it implies 'taken' as true;

"Well", says Plato in 'Republic', "but have you not noticed that opinions divorced from science are ill-favoured? At the best they are blind. Or do you conceive that those who, unaided by the pure reason [knowledge], entertain a correct opinion, are at all superior to blind men who manage to keep the straight path?".[68]

Aristotle's views on concepts of doxa and hupolepsis vary slightly from Plato's on two grounds: he thinks belief is knowledge to the extent that it is commonly believed to be true (truth entailing); however, he prefers to bring imagination into the equation on the grounds that imagination is 'ta eph'hemin', (It is up to us : conscious) whereas doxa is 'ta ouk eph'hemin', (It is not up to us : unconscious) – doxa has a truth value in that it can be true as much as it can be false (aletheuin/pseudesthai), that is, true/false. To Aristotle, in 'Posterior Analytics', belief is syllogistic in that its conclusion amounts to A is C (or C is A). To put it in other words:

P1- You like oranges

P2- I like you

Con. - So I like oranges

Belief drawn from this type of reasoning is fallacious, or at best true/false (aletheuin/pseudesthai).

In Aristotle, "EVERY syllogism proceeds by means of three terms. The aim of one, the affirmative class, is to show that C is A, because B is A and C is B; the negative syllogism has as one of its premises the proposition stating that one term is true of another, as its second that one term is not true of another". (69)

Predicating this statement Aristotle observes,"When a fallacious argument occurs in the second figure it is not possible for both the premises to be false in their entirety. When B is included in A no term can be predictable of the whole of the one and none of the other, as has been remarked above". (70)

If belief is what we know, then interpreting Aristotle, what we know is either imaginary (phantasia) or the result of our sensory perceptions, that is, the result of what we hear, or read, or smell or touch.

Aristotle maintains that, "...imagination is a different thing from both perceiving and thinking. Imagination cannot occur without perception...For it lies in our power to be affected by imagination whenever we wish...while holding belief is not up to us.... It remains, then, to see if imagination is the same as belief, as there is both a true and a false variety of belief. Belief, however, is followed by conviction, as it is not possible for those that hold beliefs not to be convinced of the things in which they believe...In other words, the conviction that accompanies all belief is produced by persuasion. These points, then, show that imagination could not be belief with perception or belief through perception or a combination of belief and perception".[71]

In Aristotelian terms there is a relative degree of happiness (highest human good or eudemonia [sometimes spelt Eudaimonia]) in human activities, deception included, and a relative degree of pleasure (of the mind) associated with them particularly in 'Restorative' activities such as lying, "... pleasures differ in kind; for those that come from noble acts are different from those that come from base ones". [72]

The pleasure that is drawn from lying, indeed the pleasure that is drawn from acts of deception even as restorative behaviour, 'concerned with intercourse in words and action' is the pleasure 'drawn from the hindrance that activities receive from the pleasure derived from other activities". [73]

Aristotle's views on deceptive behaviour, lying in particular, appear to differ from Plato's. Or do they?

Plato allows lying in certain circumstances. Aristotle, seemingly, views lying as antithesis to the components of happiness (eudemonia - the highest human good). But this is where his views on the matter lose clarity. Certain aspects of his views point to Plato's direction.

Aristotle's main concern is human nature, and as such he

admits that concepts such as the highest human good are relative to their desirability (for themselves or for other desirable goods) and the degree of Pleasure they bring to life, quantitatively or qualitatively.

Ultimately "...all men seek to obtain pleasure, because all men desire life. Life is a form of activity, and each man exercises his activity upon those objects and with those faculties which he likes the most....And the pleasure these activities perfects the activities, and therefore perfects life, which all men seek. Men have good reason therefore to pursue pleasure, since it perfects for each his life, which is desirable thing. The question whether we desire life for the sake of pleasure or pleasure for the sake of life, need not be raised for the present. In any case they appear to be inseparably united; for there is no pleasure without activity, and also no perfect activity without its pleasure. This moreover is the ground for believing that pleasures vary in specific quality. For we feel that different kinds of things must have a different sort of perfection. We see this to be so with natural organisms and productions of art, such as animals, trees, a picture, a statue, a house, a piece of furniture. [74]

One cannot but infer from the above background that although Aristotle tries to stay shy of legitimatizing lying (like Plato) somewhere in his scheme of the highest human good, he inevitably moves towards acknowledging deceptive activity in the broader package of life activities that we call (human) nature. Plants, animals and, in this case humans, use deceptive behaviour either for their own or for others' happiness. In this respect Aristotle affords lying in-moderation a type of utilitarian morality when it is used for the greater good, not so very different to how Niccolo Machiavelli (1469-1527) perceived lying for utilitarian purposes in his political treatise called 'The Prince' (1532) .

Machiavelli thinks, in effect, that if all humans were good, then there would be no concept of the good itself. There has to be some bad in order to give the good its validity, its raison d'etre; "You must realize this", says Machiavelli, "that a prince, and especially a new prince, cannot observe all those things which give men a reputation of virtue, because in order to maintain his state he is often forced to act in defiance of good faith, of charity, kindness, of religion. And so he should have a flexible disposition, varying as fortune and circumstances dictate. As I said above, he should not deviate from what is good, if that is possible, but he should know how to do evil, if that is necessary". [75]

Machiavelli maintains that humans, by and large, believe what they see; therefore appearances matter the most; "Men in general judge by their eyes", says he, "The common people are always impressed by appearances...".[76]

Machiavelli shows no hesitation to advise rulers to be cunning, to appear a man of compassion, of good faith, of integrity; appear a kind person, above all appear to be a religious person – Machiavelli cannot be more emphatic about the appearance of being religious because it is on this ground particularly that he [the ruler] will always be judged as an honourable person and will be universally praised for it.

In terms of the political philosophy, Machiavelli is famous because of his allegory of the fox and the lion. He advises rulers to act like a fox to recognize traps and like a lion that frightens off wolves; "So it follows", writes Machiavelli in 'The Prince', "that a prudent ruler cannot, and must not, honour his word when it places him at a disadvantage and when the reasons for which he made his promise no longer exist".(77)

He sees naivety in the lion which from arrogance and credulity does not see traps, unlike foxes which draw their power not from the muscles of their body, rather from their intelligence and cunningness, "...those who have known best how to imitate the fox have come off best. But one must know how to colour one's actions and to be a great liar and deceiver. Men are so simple, and so much creatures of circumstance, that the deceiver will always find someone ready to be deceived". (78)

Machiavelli does not see virtue and morality in terms of black and white by nature; in fact he constantly reminds us that such qualities are fickle and mostly demonstrate shades of both black and white.

While Machiavelli advocates hiding one's deceptive intentions behind the appearance of truth, St. Augustine of Hippo (354-430 AD) finds this act a corruption of the heart (soul) ; He " who holds one opinion in his mind and who gives expression to another through words or any other outward manifestation" [79], is a liar who has lost his heart.

St. Augustine finds lying a 'magna quaestio' deserving considerable attention, so much so that he wrote the book, 'De Mendacio' (On Lying) in 395 AD and dedicated one chapter of his Retractations (1) to an elaborate deliberation on the topic complete with supporting examples of scenarios where lying does or does not count as a sin.

A discussion of St Augustine's standing on the issue of lying needs a space of its own; what conclusion one can draw from his arguments is that it is the intention of the liar, above other considerations, that matters, "For, a person is to be judged as lying or not lying according to the intention of his own mind, not according to the truth or falsity of the matter itself". [80]

Throughout his writing St Augustine struggles to convince himself that all lying is bad lying even if it is meant to save somebody's life, "Since, therefore, eternal life is lost by lying, a lie may never be told for the preservation of the temporal life of another. In very truth, some are indignant and angry if someone is unwilling to lose his soul by telling a lie so that another may grow a little older in the flesh". [81]

However, St. Augustine, as other like-minded philosophers, eventually comes to the conclusion that human relationships are not so straightforward and such blanket rulings do not fit human life; that there may be occasions, such as issues 'concerning the worship of God' when it is a good and pious deed to speak falsely. Every lie, therefore, is a sin although the degree of it is dependent upon the intention as well as the topic of the lie, "Hence, we must consider which of the two persons lies more grievously, he who tells what is false without the intention of deceiving, or he who tells what is true in order to deceive..., because persons must be judged according to their deliberate intention". [(82)]

St Augustine is constantly emphatic about the intention of the lie that amounts to a sin, except one area where he finds lying permissible, "In the treatise I am excluding the question of jocose lies, which have never been considered as real lies, since both in the verbal expression and in the attitude of the one joking such lies are not accompanied by a very evident lack of intention to deceive, even though the person be not speaking the truth". [(83)]

Since the publication of 'De Mendacio', the views and the stories contained in it have attracted commentary from many a theologian/moral philosopher. For instance, in his treatise Saint Augustine recites the story of a St. Firmus of Tagaste who had refused to reveal to the authorities the identity and the whereabouts of a fugitive who had found refuge with him. The saint resisted exposing the man even when he (the saint) was subjected to torture. On grounds of his religious and moral steadfastness he was brought before the emperor who praised his devotion to the faith and pardoned both him and the fugitive.

Saint Augustine uses stories of this kind to justify his universal opposition to lying although, as we have seen, this claim is not exactly correct particularly when measured against interpretation of such claims by other influential theologian/ moral philosophers, notably Thomas Aquinas (1225-1274).

Aquinas, in his unfinished work 'Summa Theologica' (1485), while seemingly agreeing with Augustine that one should not tell a lie to save another from danger, declares it lawful to hide the truth in some circumstances, 'prudently'. Aquinas pretends that this is what Augustine had meant (Summa Theologica 11:110:3), although Aquinas would have recognized that in the final analysis not telling the truth amounts to telling a lie.

The French philosopher of the Renaissance era Michel de Montaine (1533-1592) distinguishes between hiding the truth and telling lies (Essais, 1850). He finds hiding the truth a worse act of deception than straight lies in that a straight lie has 'one face' whereas hiding the truth leaves the case open to interpretation and contradictory understanding which potentially can lead to contradictory actions.

However, even though Montaine finds lying a lesser evil, he does not underestimate the gravity of lying and its effects on relationships. In fact he makes it a conditional distinction suggesting 'if' falsehood had one face like the truth, then it would be a lesser evil compared with the hiding of the truth because we would then be able to counter-position ourselves in response to it. [84]

Where Montaine seems somewhat different to other philosophers who have written on the subject is in accepting the inevitability of lying since it 'grows with the child's growth' and once the tongue has reached an age that tastes its flavours, it is difficult to imagine avoiding it (they cannot even if they tried since lying is a sign of mastering some essential cognitive skills), "Lying…grow[s] with a child's growth, and once the tongue has got the knack of lying, it is difficult to imagine how impossible it is to correct it". [85]

Perhaps the most controversial views on the subject of lying versus telling the truth comes from the German philosopher Immanuel Kant (1724-1804).

It is generally agreed by Kantian commentators that he is uncompromisingly opposed to deceitful action. That said, such categorical statement about any of Kant's philosophical theories is difficult to substantiate. Kant is a particularly difficult philosopher to follow in all respects. In relation to lying, Kant has allocated a great deal of his complex book, 'Critique of Pure Reason' (1781/89) as well as numerous essays to the subject, among them the most famous short essay written in 1797 titled 'On a Supposed Right to Lie from Philanthropy' - in response to an article by the French political philosopher, Henri-Benjamin Constant (1767-1830), in the 'periodical France', Part V1, N0.1, P.123, where Kant mentions the case known as 'The Murderer at the Door'. Under this concept, Kant says, if a would-be murderer comes to your door and asks whether a person (your friend) whom he intends to kill is taking refuge in your house, you have the duty to tell him/ her the truth. If you did not tell the truth, you abrogated your moral, ethical and juristic duty.

In his article Constant had almost ridiculed Kant for his theories on the concept of lying, claiming that if such a moral duty existed, as Kant was advocating - that all men tell the truth all the time - strictly and unconditionally, almost amounting it to a crime , it would be impossible for the society to carry out its normal functions.

Constant further qualified his views saying that concepts of duty and RIGHT (in both sense of the word) are inseparable.

One has duty towards another when that person has the right to know the truth; where there are no rights there are no duties. Applying Constant's theory to the case of the would-be killer, no right (moral, ethical, juristic) to know the truth existed.

The truth is that getting a clear, simple definition of lying from

Kant is not possible because he defines lying (jdm LÜgen erzählen, to tell lies) a wrong in various contexts namely: ethical (falsiloquium dolosum), juristic/legal (falsiloquium dolosum in praejudicium alterius) and in the context of right (falsiloquium dolosum in praejudicium humanitatis).

Here I am inclined to differentiate between 'right' and 'RIGHT', the former to suggest 'truth, honesty, correctness' and the latter to mean 'entitlement'. If ever there were a generic, simple definition, it would be the one suggested by Allen W. Wood of Stanford University, USA (Eidos 2011) to the effect that lying is an intentionally untruthful statement that is contrary to duty, especially contrary to a duty of right.

The other words in [my paraphrased version of] Wood's Kantian definition, also Kant's favourite, are intention and duty to which we shall return.

By way of qualifying, rather expanding on, the above definition, Kant, as a point of departure, categorizes lying into (a) dissimulation-(concealment of one's thoughts and feelings as in the answer that we give to the question: HOW OLD ARE YOU?), (b) silence - which may make an untruth be interpreted as true and (c) lying per se - the truth suppressed and replaced by substitutes (falsiloquium dolosum).

The other point to add is that Kant recognizes that sometimes we all do lie to ourselves – inner lies – as if we were somebody else, for whatever the underlying reason or reasons.

As long as we recognize that our inner thoughts and statements are untrue, and under the guardianship of our conscience, we are not motivated by them to externalize them so as others believe them to be true, then they fall under the category of hypothetical principle.

Kant makes hypothetical (conditional) principle, under the guardianship of our conscience, permissible because sometimes circumstantial utilitarianism such as social manners, etiquette and the like require telling untruth - notations such as 'Sincerely yours', 'Truly yours' or complementing someone when the expression carries the meaning of sociability only; or hide the news of a terminal illness from a person.

Beyond these words, in order to understand Kant's philosophy on 'lies' and acts of lying, there are a number of other terminologies that require familiarity, most importantly 'Categorical Imperatives', but also concepts of Hypothetical Imperatives, Moral Imperatives, maxim, a priori, a posteriori and a number of others.

Categorical imperative commits humans to never tell untruth because by doing so they negate (break the promise inherent in the Ten Commandments) the universality of the Biblical 'Golden Rule'- Mathew 7:12 "So in everything, do unto others what you would have them do unto you", a reflection of the

Law of Moses: Leviticus 19:18 of the Jewish scriptures.

Kant distinguishes the Categorical imperative from the Hypothetical imperative in the sense that the latter, conditionally allows doing things so long as they do not impinge on the rights of others who are free to accept or reject propositions and/or consequences.

Categorical imperative also entails a Maxim – the unnegotiable, fundamental moral principles, whether of objective or subjective nature, that bind humans to being (universally) truthful to each other – a bondage that is based on truth even for utilitarian purposes.

Categorical imperative does not allow generalization of personally formulated maxims such as 'I can do such and such because others do it'. Kant makes this comment to indicate that by universalism he does not mean generalization of the sort.

Kant regards humanity an entity by itself wherein rests the supreme principle of universal morality necessary for its soul (a priori). Our duty is to adhere to the moral rules contained in our soul – our attachment to human dignity. On this account lying has no place.

Experience, on the other hand, is a posteriori - secondary in respect of moral laws. Experience makes us good or bad according to our intentions.

It seems to me that the more Kant tries to clarify his philosophy on lying, the more contradictory his views become and the more confusing. Imagine this scenario:

The soldier at the war front has the duty (militarily) to kill the enemy soldier he/she confronts (and vice versa). In the circumstance (conditional) he/she has the RIGHT (entitlement) to defending himself/herself and, by extension, his/her country. But killing another human (they don't know anything about each other except that each knows the other is there with the intention to kill him/her) is not right measured against the universal principle of moral law but in accord with the biblical statement 'So in everything do unto others what you would have them do unto you'. Killing each other at the war front is a utilitarian duty even though it contravenes Kant's universal natural law. The word 'natural' itself is contentious. In the world of animals, to which humans supposedly belong, killing is natural – wrong but a priori.

We come to the end of this chapter acknowledging that lying, concepts of deception generally, have occupied many a philosopher's mind over centuries, and they will continue to do so as the next chapter – in some ways a continuation of the current discussions - will demonstrate.

distinguish
cause
deduce
evaluate
rely on
contrast
ascertain
count on
effect
discern
rationalize
analyze
separate
manufacture
quantify
scrutinize
appraise
mechanize
examine
calculate
compare
measure
transcend
sense
attune
dream
empty
unify
embody
inhabit
enliven
evolve
trust
flow
disown
play with
dissolve
intuit
risk
question
inquire
feel
inspire
reveal
try on

"Lying is a complex activity, one that we often blame, despite the fact that several times it may be the best ethical option left to us. While lying can be seen as a threat to civil society, there seem to be several instances in which lying seems the most intuitively moral option. Besides, if a sufficiently broad definition of "lying" is adopted, it seems utterly impossible to escape lies, either because of instances of self-deception or because of the social construction of our persona."

Andrea Borghini
ThoughtCo

Chapter Six

Moral and Ethical Rudiments

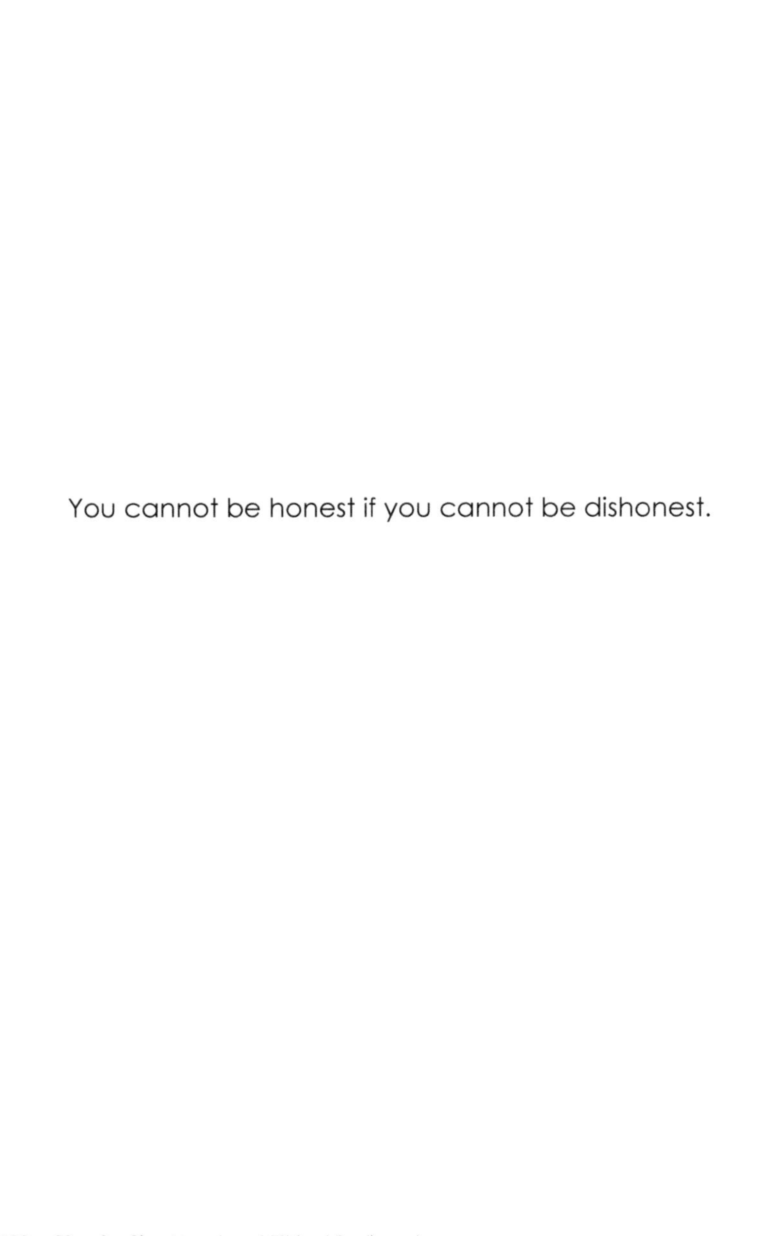
You cannot be honest if you cannot be dishonest.

Moral and Ethical Rudiments.

The title of this chapter is suggestive of the monozygotic nature of morality and ethics - the common view that the two principles are conjoined twins (or even one and the same).

I begin this chapter with the premise that contrary to this view the two precepts, complementary as they may be, ought to be distinguished from each other by virtue of their differing geneses. This distinction warrants being established before we begin to apply each precept to determining 'right or wrong' of lying.

According to the ancient Chinese Taoist principle of Yin and Yang (4th-3rd century BC) morality is defined as being a Being - the essence innate in human nature for assessing good from bad values, 'jugement de valeur' -judgement of values (the translation into English of this phrase as 'value judgement' is confusing). Thomas Hobbes, to whom I shall later return, calls this essence 'foro interno' (Leviathan,1651). Ethics, which Hobbes calls 'foro externo', is external standards for Doing - actions driven by judgements based on non-innate standards - 'jugement de faire'.

Morality is that force within - that which guides one's relationship with oneself and, from there, with the outside world (In > Out). Ethics, on the other hand, consist of historical and sociocultural precepts guiding inter-relationships in a variety of contexts: sociological, theological, economic, political, legal (civil or otherwise). These contexts feed customary standards. I conclude, therefore, that the reciprocal relationship between morality and ethics is that of nature and nurture.

I use the word 'nature' to include within its definition the dichotomy of pleasure and pain, the dual principal terms used ubiquitously by moral and ethical philosophers to signify comfort and happiness viz-a-viz suffering and unhappiness.

Nurture, on the other hand, is defined as our personal life experiences as well as the historical experiences of others conferred onto us to enhance our chances of survival (comfort) in a given cultural group.

The desire for pleasure (avoidance of pain) makes our behavioural responses conditional and circumstantial. It constantly puts us at the crossroads of right and wrong in innumerable situations, predictable as well as unpredictable, innocently or by design – hence volumes of philosophical discussions on the subject.

The philosophy of Yin and Yang in Taoism, for instance, holds that existence is based on contrasting forces that although independent are interdependent – Yin is Receptive while Yang is Active. The universe organises itself into dual cycles that form differences: cosmological, anthropological, sociological, psychological, technological, scientific.

Indeed such distinctions are apparent in Chinese philosophies prior to Taoism. In Confucianism, for instance (6th-5th century B.C.), morality and ethics are in a hierarchical relationship whereby the former is considered 'Majestic' and the latter 'Real' , the two entities work hand in glove (distinction intended) to attain happiness (pleasure) for the individual and justice, compassion, honesty, mirth, reverence and strength (collective pleasure) for the society.

As a matter of fact, Confucianism (and later Neo-Confucianism) views morality in proportional as well as hierarchical scales in relation to one's obligations towards oneself, towards one's family, then friends/acquaintances and, finally, towards strangers, acknowledging that one cannot remain moral in all respects to all people all the time.

This, so called Relational or Situational definition of morality (intuitive morality) puts us in counter-path to Kant's principles of absolutism and universalism about which we have made points in the previous chapter and to which we shall return later in this chapter.

Almost at the same time that Confucianism was inspiring moral behaviour in (the greater) China, in ancient Persia, Zoroastrian philosophy, founded by the Persian speaking Zarathustra (6th century B.C.), came to reconcile the two concepts by putting morality and ethics in a reciprocal, interdependent relationship.

Zoroastrianism is centred upon a dualistic cosmology of Ahuramazda (Lord of Light- good spirit, truth and justice - Pleasure) in constant struggle with Ahriman (Lord of Darkness, untruth, chaos, confusion, disappointment and strife - Pain).

By adopting the three tenets of Asha - three commandments: Hu-mata (good thoughts), Hu-khta (good words) and Hu-varshta (good deeds), individuals path themselves along a mutually respectful relationship with the universe (including animals, plants, etc., all of which have evolved from one essence – LIGHT the prefix 'Hu' means essence or nectar).

The three tenets are guidelines to secure happiness and prevent mental and physical discomfort. Zoroastrian doctrines are amongst the first, arguably the first, to condense concepts of morality and ethics into monotheism (one God). Zarathustra's Asha reconciles a number of disparate theories that we ascribe to moral (In > Out) as well as to ethical (Out > In) virtues.

The first principle – good thoughts – calls on the individual to source their own feelings and emotions, their imagination and above all their memory of personal experiences to think about their relationship with others – these resources shall invite LIGHT - majestic thinking - into one's mind with which one can (identify and) suppress the Evil in oneself (awareness of good and bad thoughts); thus allowing the Good (Good and Evil co-exist in human nature) to guide their course of action. In modern Persian the word 'pendaar' is translatable into the English word 'ponder' and both signify the liquidity of thoughts and the process of thinking.

Asha's second principle refers to 'word' or 'vac' which means voice (pleasant Vs painful language, awareness of good and bad words), signifying malleability in the nature of voice (words) – we need to remind ourselves here that Zarathustra has a practical view of language unlike theologians like St. Augustine who viewed language a gift from God to humans so that they could inter-communicate their Christian thoughts; and if they used their language ability for deceptive purposes they would have committed a sin (see chapter 5).

Talking about words, Joseph Poulshock, in his doctorate dissertation (2006) titled 'Language and Morality: Evolution, Altruism, and Linguistic Moral Mechanisms' asks, "Does morality require language? Could humans have developed a morality without language? Could we have developed a language without morality? What is the evolutionary relationship between language, altruism and morality? These language-related questions connect to the general Darwinian problem of altruism...." [86]

Poulshock, on the basis of research by Hamilton (1964), Dawkins (1976) and others, finds that altruism as manifestation of morality in the context of evolution is hierarchical and scaled, reminiscent of what was mentioned before in relation to Confucianism. In other words, altruism, ie., selfless concern for the welfare of others, in humans and other species, in the context of evolution, is more a matter of appearance than genuine behaviour since organisms are primarily interested in self reproduction and communal relationship through kinship rather than through non-relatives.

There is, according to Poulschoch, based on research by the American biologist Robert Trivers (1971), a form of apparent altruism directed toward non-relatives which Trivers named 'reciprocal altruism'. Reciprocal altruism encompasses principles of morality and ethics as it prescribes mutual obligation between the individual and the select group (SG) – a give and take arrangement.

To return to the questions posed by Poulshock and our discussion on Zarathustra's second maxim – Good Words, we established in the previous chapters that the ability to use language is embedded in human evolution. Language is the tool with which we think and with which we relate to others and others to us in reciprocal rapport; as Ludwig Wittgenstein said, "we dwell in language". [87]

In Zarathustra words facilitate our transition to varshta or varesh or the process of becoming – aspiring to be. Words, therefore, are essences (the prefix hu preceding each maxim signifies essence or nectar) that shape our destiny; they are staota – that which brings into existence. The tripartite essences, ie., thinking, voicing and becoming, for the virtuous are the path to happiness – Garo Demanae (the Grand Eudaimonia: the celestial house of Ahuramazda, the house of beauty, music and songs, the place where the union of oneself with others (and things) happens as reflected in the poetic myth, 'The Conference of Birds' by the Persian poet Farid ud-Din Attar (1145-1220).

The allegory of 'The Conference of Birds' is the story of the birds of the world gathering in a conference to elect a leader that would lead their way to 'Truth'. After much deliberation the Wise Hoopoe bird [(88)] is elected the leader.

Hoopoe tells the birds that the truth they are seeking is called 'Simorgh' – Phoenix like (Phoenix is a bird that self regenerates; it gets cyclically reborn), and warns the birds that the route to Simorgh is treacherous and the journey torturous and that it is divided into stages of ascension.

From the travelling party of thousands, thirty birds make it to the destination. The birds, having travelled considerable distances along seven stages, each posing physical and mental pain, numerous challenges, arrive at the destination only to find themselves as one; they discovered that Simorgh - Si means thirty and morgh means bird, hence Simorgh or thirty birds - was a reflection of their own faces in a lake - the lake in which the birds see themselves is an Implicit lake. Mathematically the word 'implicit' signifies an expression in which the dependent and the independent variables are not on the opposite sides of an equation. Further, semiotically speaking, words are transformational and generative, to borrow Chomsky's terminology. In Zarathustra, words, thoughts and actions are also transformational and generative. Readers may wish to refer to Noam Chomsky's Transformational Generative Grammar.

One might infer from this mythical story that morality is the ultimate Goodness embedded in the dissolution of differences, the abdication of self(ishness), in seeing oneself as if Other(s) where self merges into a collective oneness – god – hence the foundation of 'monotheism' by Zarathustra, as far as we know.

The allegory is reminiscent of the video clip of an experiment reportedly conducted by a philosophy lecturer in America.

At the end of a session on morality the lecturer poses the question to the attending students 'who they consider to be or to have been a good person and who the bad?'

Students volunteer different names.

The lecturer, holding a cardboard gift box the size of a square Kleenex facial tissue box with a lid, announces that he wants to get the box he is holding in his hands passed on to everybody in the class, one person at a time. No one is to look inside the box except for the person in possession of the box.

He then gives the box to the first student in the front row. The student lifts the lid, looks inside it, closes the lid and passes it on to the next student, and the next till it reaches the last student.

While she (the last person in the class) is looking inside the box the camera pans in, as if her eyes and ours, to show a mirror attached to the bottom of the box and nothing else.

Each student had seen her/his own face in the mirror.

Attar's allegory and the philosophy lecturer's experiment are acknowledgement that good and bad (pleasure and pain) co-exist in each individual. It is the motivation triggered by circumstances that drives the person towards one or the other quality enacted in words that they say to themselves and to others, the way they behave in respect of themselves and of others. The consequences of what and how they say and act maximizes or minimizes pain;

"Nature", wrote Jeremy Bentham (1748-1832), an advocate for the philosophy of Utilitarianism, "has placed mankind under the governance of two sovereign masters, pain and pleasure. It is for them alone to point out what we ought to do, as well as to determine what we shall do. On the one hand the standard of right and wrong, on the other the chain of causes and effects, are fastened to their throne. They govern us in all we do, in all we say, in all we think...". (89)

Like Zarathustra's good and evil, Bentham regards pleasure "the only good" and pain "without exception, the only evil", (1970, 100) - they alone count as the moral or the immoral consequences of an act.

Our ability to reason (think) is as much capable of motivating us to do good (cause comfort/pleasure) as to do evil (cause pain); however, to Bentham it is the consequences of the act, not the motivation/s behind the act, that matters the most. One may have had the motivation to deceive someone in a financial transaction but if the expected consequence did not materialize, or it resulted in favour of the victim, then 'good', not 'bad' has resulted. In such a case, according to Bentham, whether the intention (motivation) was moral or immoral is inconsequential. [we may look at this issue in the context of an event in America where a white police officer who allegedly suffocated an unarmed African American man to death by pressing his knee against his airways, ignoring his plea 'I can't breathe' recorded on camera, is, allegedly, charged with 3rd degree murder and/or man-slaughter and not charged with 1st degree murder on grounds that the officer did not intend to kill (lack of motivation to kill). Many observers find the charge too lenient. They believe the officer was motivated by racism and view the man's death the consequence of the officer's deliberate actions calmly kneeling on the victim's neck ignoring his perfectly clear and audible plea].

Bentham also asks us to be mindful of the fact that acts such as lying to others generally happen spontaneously and as such it is normally not possible for the person to engage in calculating the consequences of their lying.

By thus moderating our judgement of good and bad, Bentham makes the morality of acts such as lying causative, relative and circumstantial.

Bentham's views on the morality of lying, along with views of other utilitarian consequentialists like John Stuart Mill and Henry Sidgwick on the subject, differ significantly from deontologists like Immanuel Kant who declared that, "To be truthful (honest) in all declarations, is, therefore, a sacred and unconditionally commanding law of reason that admits of no expediency whatsoever." [(90)]

Both Mill (1806-1873) and Sidgwick (1838-1900) expanded on the views expressed by Bentham in respect of utilitarianism of human thought and action.

Mill, in his essay 'Utilitarianism' (1861), made the point that the aim of good action should be maximization of happiness (pleasure viz.a.viz pain) but he put a caveat in that assertion by scaling the quality of pleasure (and pain) from high to low, remarking that one would be better be Socrates and dissatisfied than a fool and satisfied. He found that utilitarian morality was well placed to accommodate and enact honesty, truthfulness and justice. He also expressed similar views to Bentham's that people are not in a position, nor should they force themselves to be, to calculate whether a particular action they are performing will maximize pleasure or not, rather they be guided by the general rule that honesty, truth and justice are inherently pleasure invoking principles.

Sidgwick went even further in simplifying the concept of morality, calling it 'intuitive' or 'common sense morality' (Methods of Ethics, 1874). He thought anyone would know that honesty, kindness, compassion, empathy, justice, gratitude, truthfulness, etc., did not need systematic, philosophical debate to be proven pleasurable. They were self-evident principles that we all recognize and appreciate when we are their recipient and feel pain when we experience dishonesty, ingratitude and injustice. He went on to say that even these self-evident principles can be at times in conflict with one another and unclear in their application unless they are put into a coherent, well defined system of morality that identifies the circumstances and the degree to which each circumstance can produce pleasure or displeasure – a system of co-ordination and sub-ordination. [(91)]

Since this approach is not easy, maybe not possible, given that people are different - they live in different circumstances, they have needs that cannot be mapped nor predicted, their motivations change as they progress through life- it is best to leave judging goodness or badness of thought, language and action, including in relation to deception, to the realms of common sense.

Concepts of pleasure for virtue and pain for vice date back at least to the 4th century B.C. Greek philosopher Aristippus (and later Epicurus although from a different perspective) who is credited for founding the moral philosophy that became known as 'Hedonism' or the theory of value (phrased 'jugement de valeur' in the opening paragraph of this chapter).

Although different branches of Hedonism have sprung since its historical inception (I shall refer to them shortly), they all share the theory that pleasure is the only intrinsically valuable quality for all people at all times and pain is intrinsically the contrary.

Aristippus (435-356 B.C) was a firm believer in the Socratic view that happiness is a result of moral action derived from human inner sense of right and wrong although he went one step further suggesting that pleasure, the ultimate good, is more desirable when it comes as immediate gratification rather than achieved over a period of time.

Further, Aristippus distinguished intrinsic value from instrumental value to which lying, rather deceptive behaviour generally, belongs - and so does money, for instance. Money gives us pleasure when its value affords us the ability to purchase something. The broader interpretation of the terms 'intrinsic' and 'instrumental' is that the former refers to morality (foro interno) and the latter to ethics (foro externo).

Since the inception of the theory of Hedonism, different perspectives on the concept have developed amongst them Normative and Motivational.

Normative Hedonism takes it upon itself to explain when, how and why an action, verbal, nonverbal or a combination such as an undercover police officer pretending to buy narcotic drugs from an illegal drug dealer, can be morally right or wrong, permissible or impermissible. Normative Hedonism looks at situations from two perspectives: Egoism and Utilitarianism.

Egoistic Hedonism maintains that people act to suit their own best interest regardless of the consequences that their actions may produce. Consequences matter only when they are favourable to the person performing the act and no other. Egoistic Hedonism does away with concepts of guilt, empathy and moral compassion in order to achieve certain goals. In this sense we are reminded of a statement attributed to Plato to the effect that people never do wrong, because they do what they think is right to do.

Utilitarian Hedonism, on the other hand, considers the good of a given community or communities as paramount even if morally wrong acts are committed to achieve it. Examples of such philosophy can be found in policies of colonialism, post-colonialism and the like according to which colonizing powers, by means of lying and in deceptive guises, exploit resources of other nations to the benefit (happiness; pleasure) of their own – the invasion of Iraq by a coalition of Western nations in 2003 was grounded in utilitarian lies (readers may wish to refer to a book by David Corn, titled 'The Lies of George W. Bush published in 2003 by Crown Publishers).

Not dissimilar to the doctrine of Utilitarian Hedonism is Jean-Jacques Roussseau's notion of 'amour propre' [92] (desire of self-preservation) – the notion that humans, in the process of their social interactions, whether amongst individuals or between nations, have the tendency to compare themselves with others.

Rousseau (1712-1778) believes that human social relations envelope self-improvement, private ownership, labour and a whole range of opportunities that determine their happiness or unhappiness. In such circumstance sense of competition drives humans to want to dominate others by whatever means possible, not excluding deceit – back to Darwin's theory of the Survival of the Fittest (incidentally Rousseau too sees inequalities primarily related to the inequalities that exist in the natural world including the differences in the physical strength of animals and humans).

In his famous book, 'The Social Contract' (1762), under the heading, 'The Right of the Strongest' Rousseau writes, "The strongest is never strong enough to be always the master, unless he transforms strength into right, and obedience into duty. Hence the right of the strongest, which though to all seeming meant ironically, is really laid down as a fundamental principle." (93)

While Rousseau, like Darwin, recognises survival as the fundamental natural order, he prefers to distinguish between the approach the primitive man took to secure survival and the methods that the social (civilised) man employs to that end. In this regard he takes issue with ideas on human nature expressed by some important moral philosophers like Thomas Hobbes to whom I shall return after a brief note on Motivation Hedonism.

Rousseau's view is that 'men in a state of nature' are neither virtuous nor vicious since they are driven by 'amour propre' and should not be judged 'wicked' "for the mischiefs they do one another" (94), and continues, "It appears, at first view, that men in a state of nature, having no moral relations or determinate obligations one with another, could not be either good or bad, virtuous or vicious; unless we take these terms in a physical sense, and call, in an individual, those qualities vices which may be injurious to his preservation, and those virtues which contribute to it; in which case, he would have to be accounted most virtuous, who put least check on the pure impulses of nature." (95)

As prelude to my discussion of Thomas Hobbes' views on human nature, I am tempted to refer to the comments that Rousseau makes on Hobbes, suggesting that, "Above all, let us not conclude, with Hobbes, that because man has no idea of goodness, he must be naturally wicked; that he is vicious because he does not know virtue; that he always refuses to do his fellow-creatures services which he does not think they have the right to demand; or that by virtue of the right he justly claims to all he needs, he foolishly imagines himself the sole proprietor of the whole universe..., he [Hobbes] ought to have said that the state of nature, being that in which the care for our own preservation is the least prejudicial to that of others, was consequently the best calculated to promote peace, and the most suitable for mankind." (96)

Motivational Hedonism - which incidentally reminds us of ideas put forward by the psychoanalyst Sigmund Freud, philosopher Epicurus, biologist Charles Darwin and particularly the Utilitarian philosophers of 18th /19th centuries discussed before: Jeremy Bentham, Henry Sidgwick and John Stuart Mill - the desire to experience pleasure (gain benefit) and avoid pain (lose Benefit) is solely responsible for motivating human action, verbal or nonverbal, conscious or unconscious.

Any discussion of human motivation and human desire for pleasure as a moral principle is incomplete without a look at the moral philosophy of Thomas Hobbes (1588-1679) whose views on human tendencies, including the tendency to lie and deceive, have been the source of inspiration for all who have since ventured into the subject.

Hobbes had a practical view of human nature, and of sense of morality in it. The fundamental tenet of Hobbes' views is that humans are motivated by self-interest – naturally; "...every man is presumed to do all things in order to his own benefit, no man is a fit Arbitrator in his own cause....". [97]

Hobbes prefers to take us to the opening chapters of human life on earth, before the existence of civil society, wherein academic conceptions such as morality and ethics, even concepts about god and the like, were not reflected in the brute reality of survival; man's sense of reason - in the service of his other senses: olfactory (smell), auditory, etc., was there to minimize pain and maximize pleasure – the rational pursuit of self-interest;

Hobbes wrote, "Pleasure, therefore, (or Delight) is the appearance, or sense of Good; and Molestation or Displeasure, the appearance, or sense of Evill. And consequently all Appetite, Desire, and Love, is accompanied with some Delight more or less; and all Hatred, and Aversion, with more or less Displeasure and Offence. Of pleasures, or Delights, some arise from the sense .... Of this kind are all Orientations and Exonerations of the body; as also all that is pleasant, in the Sight, Hearing, Smell, Tast, or Touch; Others arise from the Expectation, that proceeds from foresight of the end, or Consequence of things; whether those things in the sense Please or Displease: And these are Pleasures of the Mind of him that draweth those consequences; and are generally called JOY." [98]

Such 'joy' derived from the desire for pleasure is reflected in Hobbes' own experience anecdotally told about him that once passing St Paul's Cathedral in London Hobbes accosted a beggar sitting outside. A clergyman in presence saw him giving the beggar some money. By seemingly trying to score a point, or by some spontaneous, ad hoc response, he asked Hobbes whether he would have given the money had Christ not urged giving the poor. Hobbes, in reply said that he gave the money because it pleased him to see the poor man pleased (morality not ethics).

Hobbes' moral philosophy in the context of 'Lawes of Nature' borders on Darwin's theory of 'the survival of the fittest', whereby man, the thinking ape, finds a thing good if it is the object of his appetite, self-preservation and desire for survival. If man acted otherwise, that is, acted modestly and honestly, "...and perform all he promises, in such time, and place, where no man els should do so, should but make himselfe a prey to others, and procure his own certain ruine, contrary to the ground of all Lawes of Nature, which tend to Natures preservation." [99]

Life, according to Hobbes, in its purest form, is a matter of peace and war – more war than peace since peace is only attainable through social contract (ethics) whereby individuals agree to relinquish their self-interest and abide by the hypothesis, rather idealism of, "Do not that to another, which thou wouldest not have done to thy selfe". [100]

This illusion is contrary to the laws of nature built on self-interest and is fitting of religious morality which Hobbes tries to keep a distance from – particularly in relation to lying and deceptive action – an action that Hobbes finds justified morally, albeit conditionally, so "That which taketh away the reputation of sincerity, is the doing, or saying of such things, as appear to be signes, that what they require other men to believe, is not believed by themselves; all which doings, or sayings are therefore called Scandalous, because they be stumbling blocks, that make men to fall in the way of Religion...." [101]

Hobbes uses the premise of 'conditionality' liberally since, in his view, drawing conclusions as facts and absolutes in relation to anything, including lying, is no more than an academic 'discourse', "No Discourse whatsoever, can End in absolute knowledge of fact, past, or to come. For, as for knowledge of Fact, it is originally, sense; and ever after Memory. And for the knowledge of consequence, which I have said before is called science,it is not Absolute, but conditionall. No man can know by Discourse, that this, or that, is, has been, or will be; which is to know absolutely: but onely, that if This be, That is., if This has been, if This shall be, That shall be: which is to know conditionally; and that not the consequence of one thing to another;...." [102

David Hume, the Scottish philosopher, (1711-1776), also points to the direction of emotions and feelings (sentiments) as the cause of our moral or immoral actions. To him any (mental) action that produces pleasure in others is taken as virtue (rushing to save someone in house fire, for example) and if it does not, it is vice.

According to Hume, if a person knows what is right, eg., telling the truth, but does what is wrong, that is, lies, he/she simply has no desire to do the right thing which has nothing to do with rationality.

In his famous book 'A Treatise of Human Nature' (1739-1740), Hume concludes that our actions are driven by the impressions that our senses create.

In book iii of his 'Treatise', Hume deals with moral issues making a clear distinction between the impression of virtue (pleasure) and the impression of vice (pain) and concludes that moral impressions result from human action only.

Hume regards empathy, or lack of it, the foundation of what is called morality, or else immorality; "It is obvious", Hume writes, "that when we have the prospect of pain and pleasure...we feel a consequent emotion of aversion or propensity, and are carried to avoid or embrace what will give us this uneasiness or satisfaction...It is from the prospect of pain and pleasure that the aversion or propensity arises towards any object: and these emotions extend themselves to the causes and effects of that object, as they are pointed out to us by reason and experience." (103)

Hume does not share Hobbes' views on self-interest being the fundamental motivation in human behaviour (deceptive behaviour not excluded) but he comes very close to it by recognizing selfishness as such an impulse driving the individual's motivation. Further, he does not deny Plato's notion of innate knowledge of right and wrong, or the biological notion of instinct, but thinks those predispositions are shaped by our personal and social experiences (and influences) that originate our passions – he distinguishes between direct passions like fear, grief, desire, hope and indirect passions like love, humility, honesty, pride; "Nothing can oppose or retard the impulse of passion, but a contrary passion...Thus, it appears, that the principle which opposes our passion cannot be the same with reason, and is only called so in an improper sense." (104)

Hume finds universality in human nature and the impulses that drive passions, not the least in relation to moral judgements, "It is universally acknowledged that there is a great uniformity among the actions of men, in all nations and ages, and that human nature remains still the same in its principles and operations...." (105)

Hume's views on 'universality of human nature' puts us somewhat in counter-course with the other 17th century famous British philosopher, John Locke (1632-1704) who views our moral knowledge, indeed all knowledge, being acquisitive rather than 'innate'.

Locke thought of human mind as 'white paper' on which impressions of our experiences are recorded; "Let us then suppose the mind to be, as we say, white paper, void of all characters, without any ideas; how comes it to be furnished? Whence comes it by that vast store which the busy and boundless fancy of man has painted on it with an almost endless variety? Whence has it all the materials of reason and knowledge? To this I answer, in one word, from experience." [106]

Reading Hobbes' conditional morality and Hume's views on universal human nature, including selfishness, we begin to wonder whether they should not in fact be regarded Visceral Moralists.

Visceral morality theories differ considerably from Kant's deontological, duty bound, morality. Visceral morality explains morality, not in the context of Kantian meta-ethical rationalism enlightened by reason, not even in the context of the normative ethics of 'what should I do?', rather in the context of circumstances and our reactions to the stimuli and opportunities in a given situation. Visceralism even accommodates the behaviour of our ancient primitive ancestors whose raw emotions and raw genetic behavioural expressions were circumstantial and a response to their survival needs.

In the context of Visceral morality, lying, for instance, is neither moral nor immoral; not wrong not right. It is simply there – an aspect of evolution not only in terms of biology but also in terms of social and cultural phenomena. Basketing lying in vicious (anti-virtuous) absolutism as purported by Kant, for example, ignores it being circumstantial, relative and above all dichotomous and coexistent as are all matters of objective and subjective nature – the genesis of existence is based on dichotomous,dualistic contrasts: male#female, day#night, honesty#dishonesty, hot#cold.

For the benefit of the readers who may have by-passed previous discussion on Kant we may have a brief review here.

In the context of Kantian morality, lying is wrong on the grounds that it is against human dignity. His notion of human dignity is based on the correct premise that human beings are born with an evolving sense of reason, hence their unique ability to make decisions, and therefore, make choices that lead to Eudaimonia (human flourishing) – to borrow the word from Aristotle. It is incumbent on human beings to respect the power of reason, the freedom to decide and to choose for their own flourishing as well as for the satisfaction of others.

Kant regards lying a disruption of that power; he believes it corrupts the moral algorithm intrinsic in human nature – the lie one tells contradicts one's moral worth, the value that ordains humans with godliness – man is made in God's image. Humans have the absolute duty to uphold the virtue of honesty by not using others as a means for their own selfish eudemonia at the expense of the happiness of others (Eudaimonia has also been translated as 'happiness' which in Aristotelian language means satisfaction of fortune).

Kant's absolutism seems set on virtue ethics in that he regards honesty as an absolute virtue. By definition, then, honesty should stand in line with other virtues of which compassion is one other. If this line of argument holds true, then honesty can be compromised when a case demands compassion (see 'Murderer at the door' in chapter 5) in which case lying in order to save somebody's life is justified – an action that nulls kant's (amongst others) absolutism – some might say moral determinism.

I mentioned 'virtue ethics' in the context of Kant's virtue-perfect theories on lying. I need to moderate that statement somewhat in that 'virtue ethics' looks at 'right Vs. wrong' in relative as well as developmental terms. The doctrine does state that there is an ideal degree of virtuosity to which we should endeavour to reach out for, although virtuosity is composed of a range of virtues some of which sometimes need to be compromised to some degree in favour of another or other moral considerations. When it comes to moral rationalism Kant is steadfastly uncompromising. An anecdote attributed to Kant states that Kant even regards the kindness of a shopkeeper to his/her customer not out of the shopkeeper being a genuinely kind person rather out of self-interest, and logically speaking, immorality (instrumental kindness).

Visceral morality is a great deal more understanding of human psychology than Kant's views on the subject, not the least because Kant's views weigh heavily towards rationality at the expense of emotions which, all things considered, dominate rationality.

Indeed, philosophers like Alfred Jules Ayer (1910-1989) and later Charles Stevenson (1908- 19790) went as far as establishing the school of thought known as moral Emotivism.

Ayer in his book 'Language, Truth and Logic'(1936) and Stevenson in his 'Ethics and Language' (1945) expressed doubt on moral judgements recognized as logical truths or, for that matter, statements of fact.

In Ayer's view morality is no more than emotional expressions and as such whether they are considered good or bad, true or false, is in the eye of the beholder.

Ayer particularly, but Stevenson as well, had linguistic background, so they paid attention to how language can shape moral or immoral intentions, produce moral judgement, or at least greatly impact consequences that some may consider good or bad due to their own linguistic norms and habits.

The Emotivists (namely Ayer and Stevenson) believed that humans show strong emotions when confronted with attitudes or behaviour that they morally disapprove whereas in non-emotive situations, when discussing taste for example, they simply disagree.

Although Ayer and his successors were accused of subjectivism when it came to discussions on morality, they (Emotivists) stood firm on their view that there are simply no irrefutable standards of right or wrong, nor can any be devised; so the criteria to assess something right or wrong are purely relative and born of our feelings.

Visceral psychological morality makes allowances for our reactions to events. If, for example, someone queue-jumps to get on the bus while we have been waiting patiently for an hour or so for the late arriving bus, we react emotionally, very strongly, physically confrontational even, particularly if we have had a bad day already in our private or professional life. Our natural reaction does not allow the luxury of thinking whether the person who has barged in has a legitimate reason to do so, such as getting to the hospital as quickly as possible because he has received a phone call informing him that his son has had an accident at school and he cannot afford to wait in the long queue for the next bus.

The same line of reasoning applies to deceptive behaviour and lying. Some people resort to deceptive behaviour, theft and other crimes as a consequence of poverty or multiple other reasons unbeknown to us. Politicians lie frequently out of self-interest but their self-interest may have components such as the security of the nation, or economic advantages to the society. The stimuli that can drive our reactions are innumerable and mostly unpredictable. Whether we have failed our moral obligation to excuse the queue-jumper, or the queue-jumper has failed the ethics of social standard by not waiting his turn in the queue are examples that make the distinction between morality and ethics opaque or as I said in the opening paragraph 'monozygotic'.

What is to appreciate about Visceral Morality is the fact that by considering human behaviour from different levels of reactivity, it helps our further understanding of the fine differences between morality and ethics.

Visceral morality regards environmental (situational) stimuli and interpersonal relations as the primary sources of our intra and inter-personal behaviour. Children lie more intuitively and freely because their emotions, urges and passions are not yet refined fully by knowledge of sociocultural standards (ethics). Even adults who may have come of age but not of cognitive (logical, intellectual) maturity, may, like children, indulge in any form of lying, eg., exaggeration, prank, faked information, etc.

The Kantian view that all humans have the ability to reason is a valid point but the ability to reason is as varied as there are people; so their sense of morality, as Hobbes has pointed out, is conditional upon variables – genetics, education and life experience that have optimizing (or not optimizing) influence on the execution of self-interest in all respects including acts of deception.

Visceral moralists believe, as do psychologists and neuroscientists, that our emotions, driven by the function of certain parts of our brain, have serious involvement in our moral judgements, more so in our intuitive, circumstantial and spontaneous judgements that may lead to an action or a reaction including telling or not telling the truth about an issue (the psychologist, Jean Piaget, for instance, whose views on deceptive behaviour amongst children I reflected in the previous chapter, advised that we remain conscious of the fact that humans perceive the world from incalculable perspectives. Talking in relation to children lying Piaget thought children become conscious of people hiding information, themselves included, from age three if not younger).

On grounds of what has been discussed, it is reasonable to suggest that biological (evolutionary) influences, primarily, and sociocultural forces, secondarily, drive our actions and reactions in given circumstances.

There may well be a circumstance where our reason dictates moral judgements that are in contradiction to the verbal or nonverbal acts that emanate from our spontaneous, intuitive response. This is particularly true in matters of communication behaviour where ethics and morality converge in manners arising from decorum and the rules of etiquette. For example, imagine a person standing still on the steps that rise up to the platform of a train station, checking emails on their phone, blocking passage of others. In response to this circumstance, a person rushing to pass by may utter words like 'EXCUSE ME!' in a tone that carries meanings such as correction of a wrong social behaviour, avoidance of a collision, diffusion of a consequence that might result from aggression.

Our cultural memory dictates that the person committing the wrong act apologizes by saying things, 'O, I'M SORRY'. These words can be taken as a token (linguistic jargon) of morality if they have sincerely voiced out that inner feeling of guilt – discernment of right and wrong; or else a discharge of ethical appearance if they are from the repository of cultural clichés to avoid unpleasant consequences.

In summary, telling lies, rather human deceptive behaviour in general, is embedded in the history of evolution to facilitate survival wherein conflict of interest is an active ingredient. Mankind's deceptive behaviour has its origins in biological evolution, in socio-historical, cognitive and in linguistic evolution.

Our sense of morality, followed progressively by ethical standards, evolved in stages of growth towards communal living which necessitated sharing natural and material resources. It is in the nature of sharing resources that concepts of self-interest, self-preservation and survival become the motivating factor to use all means, including means of deception.

I have arrived at the end of the journey of writing this book only to acknowledge that the study of deceptive behaviour in organisms will go on filling books to come as they have in innumerable books and literature already published; perhaps more so now with the internet facilities, ebooks, etc.

Whoever said there is a much shorter way to finding out that deception has its roots in evolution than reading bulky books – a trip to the television set in the living room on a variety of discovery programs that document intra and inter-species acts of deception amongst micro-organisms, plants, insects, birds and other animals (both terrestrial and aquatic), and, of course, humans.

Endnotes

1 Collodi Carlo, Pinocchio, translated by M.A. Murray, Penguin Books,2002 edtion,PP.8-9

2 Ibid,pp.64-65

3 Ibid.P.72

4 Athena Aktipis and Carlo C. Maley, published 23 October 2017 https://doi.org/10.1098/rstb.2016.0421

5 Ibid, Aktipis et al 2016

6 Ibid, Aktipis et al 2016

7 Natural selection means the ways organisms of a species adopt to survive in an environment and procreate, maintining their physical and behavioural traits. There are two major theories applicable to an understanding of natural selection- one is the theory of Descent with Modification and the other the concept of Common Descent. Descent with modification involves observable facts; that means the changes, physical or otherwise, that we see in successive generations (offspring) of a given species. For example children of a family look different and behave differently to their parents and to each other due to random genetic mutations. Common Descent on the other hand means that all living things on the planet earth are related; that they have all descended from a common ancestor. This, therefore, suggests that over many generations and through the process of descent with modification species have mutated from a single source to different looking, different behaving species. Common descent is not an observable fact simply because we cannot go back to millions of generations to observe the gradual changes that they have been through. Our knowledge of changes are derivable from fossils, genetics, comparative anatomy, mathematics, biochemistry, and the distribution of species into various geographic and environmental conditions. We know from Darwin's studies that while the snakes in one geographic location may look different or come in different shapes, sizes and behaviour, they share similarities that link them to each other. However, because of the possibilities that each geographic location has offered them and no more, the species have adapted to that environment to survive; and over time they have passed those adapted traits in which they have been naturalized to their offspring. Over generations the traits become the NATURE of the given species.

8 George Johnson, https://www.nytimes.com. science28 Jul 2015

9 Differentiation is defined as the process whereby cells follow their programmed mechanism to generate parts of a living body: leaves of trees, our skin, our lungs etc

10 George Johnson, https://www.nytimes.com. Science28 Jul 2015

11 özhan özkaya, Cheating on cheaters-exploring bacterial social interactions to manipulate bacterial pathogens, https://phys.org/news/2018-06-cheatersexploring-bacterial-social-interactions-pathogens.html, Instituto Gulbenkian de Ciencia in Portugal

12 Mendel did not use the word gene in his work. The word gene first appeared in an English medical dictionary in 1913. Mendel used the words Dominant and Recessive. He established, through his experiments with pea plants, that each seed contained a dominant and a recessive element which when combined the plants demonstrated inherited patterns in their growth. His later experiments on flowers, corn and other plant species produced similar results

13 Cartwright John, Evolution and Human Behaviour, Palgrave, Great Britain,2000, P.46

14 Ibid, Cartwright P.7

15 Darwin Charles, The Origin of Species, Penguin Classics, 1985, P.136

16 Stephen D. Hopper (Royal Botanic Gardens, Kew, Richmond, Surry UK) and Hans Lambers (University of Western Australia) article published in Trends in Plant Science Vol.xxx, No. X 2009, p. 1 (correspondence : s.hopper@kew.org) copyright 2009 Elsevier Ltd. The article is titled : Darwin as a plant scientist: a Southern Hemisphere Perspective.

17 Livingstone Smith, David, WHY WE LIE: The Evolutionary Roots of Deception and the Unconscious Mind, St Martin's press, New York, 2004,p.32

18 Ibid, Livingstone, P. 33

19 ABC Science by Ann Jones for the program Off Track posted on Google search April 1st, 2018 titled 'Flora fatale: The carnivorous plant that scared Charles Darwin'

20 Ibid., Jones,2018

21 Ibid., Jones, 2018

22 Bayard Webster, Guile and Deception: The Evolution of Animal Courtship, digital reprint of February 19,1985, section C, page 3, The New York Times.

23 Bayard Webster, Guile and Deception: The Evolution of Animal Courtship(reiterat-

ing Paul Weldon and Gordon Burghardt)), The New York Times,Feb. 19,1985, Section C. P.3 reported in www.nytimes.com/1985/2/19/science/guile-and-deception-the-evolution-of-animal-courtship.html

24 Darwin Charles The Origin of Species Penguin classic, 1985, p. 234.

25 Daniela Canestrari and team, the University of Oviedo in Spain- March 21,2014 issue of Science magazine reported by Margaret Badore – March 20,2014 under the title Cuckoos and crows teach us how parasites can be good

26 From: Parker Joseph, Myrmecophily in beetles(Coleoptera): Evolutionary patterns and biological mechanisms, Myrmecological News, 22, 2016, 65108. Retrieved from http://www.antwiki.org/wiki/images/5/52/parker- 2016. pdf

27 Darwin Charles, The Origin of Species, Penguin Classics, 1985, P. 235

28 Barash David, Sociobiology: The Whispering Within, Fontana/Collins, 1981, P.7

29 Heyes, Cecilia, Social Cognition in Primates, printed in Animal Learning and Cognition, Academic Press, SanDiego,USA,1994 edited by N.J.Mackintosh P.294

30 Helmut Sick, Rio de Janeiro, Nova Fronteira, retrieved from https://www.wikiaves.com>forum

31 Bergman Jerry, Freud and Darwinism, Journal of Creation 24(2)2010, Retrieved from https://creation.com>pdfs

32 Freud Sigmund, Introductory Lectures on Psychoanalysis, Penguin Books,1991,P.326

33 Bergman Jerry, Freud and Darwinism, Journal of Creation 24(2)2010, Retrieved from https://creation.com>pdfs

34 Hall Calvin S. and Gardner Lindzey, Theories of Personality, Third Edition, John Wiley &sons, New York,1978, P.36

35 Sigmund Freud, Beyond the Pleasure Principle, The International Psycho-analytical Library No 4,Translated by Caroline Jane Mary Hubback fromthe 2nd German Edition, revised & edited by Ernest Jones, theInternational Psycho-Analytical Press, London & Vienna, P.4

36 Carver Charles S. and Scheier Michael J., Perspectives on Personality, Fifth edition, Pearson, Boston,2004, P.196

37 Romeo Vitelli, https://www.psychologytoday.com posted November 11,2013

38 Ibid, Vitelli 2013

39 Ibid Vitelli 2013

40 Berndt Thomas J. Child Development, Harcourt Brace Jovanovich College Publish-

ers, USA,1992,PP. 298-299

41 Ibid, Berndt P. 299

42 Tuckman Bruce W., Educational Psychology, From theory to Application, Harcourt Brace Jovanovich College Publishers Fort Worth,1992, P.171

43 Berndt Thomas J. Child Development, Harcourt Brace Jovanovich College Publishers, USA, 1992, P.341

44 Ibid (Berndt) P. 341

45 Piaget Jean, The Child's Conception of the World, translated by Joan and Andrew Tomlinson, New York: Harcourt, Brace & World, Inc,1929, P 197

46 Saul McLeod, https://www.simplypsychology.org, 2018

47 Gleason Jean Berko, The Development of Language, Charles E. Merrill Pub. Company, Ohio US, 1985 P.7

48 Krebs j. R. and Dawkins, R. Animal signals: Mindreading and manipulation, retrieved from Behavioural Ecology, An Evolutionary Approach, 2nd Edn, edited by Krebs and Davies, Blackwell Scientific publications Oxford, 380402. Google Scholar

49 Barrett, Paul et al, Charles Darwin's Notebooks, 1836-1841. Cambridge University Press and British Museum – Natural History, 1987

50 Kerr Philip, ed., The Penguin Book of Lies, Penguin Books,1990, P. 3Kaplan David and Manners Robert A. Cultural theory, Prentice-Hall, Inc, New Jersey, 1972, PP. 147-150

51 Kaplan David and Manners Robert A. Cultural theory, Prentice-Hall, Inc, New Jersey, 1972, PP. 147-150

52 Homo Erectus: Methods of communication- UWC (planet.uwc.ac.za > loe > page-314)

53 Kottak Conrad Phillip, Anthropology: The Exploration of Human Diversity, Random House, Inc, New York, 1974, P. 251.

54 Ibid, Kottak, 1974

55 Scott-Phillips Thomas C., Evolutionary Psychology and the Origins of Language, Editorial for the special issue of Journal of Evolutionary Psychology on the evolution of language, JEP,8,April 2010, P.294

56 Oesch Nathan, Deception as a Derived Function of language, University of Oxford posted September 27, 2016 retrieved from https://doi.org/10.3389/fpsyg.2016.01485.

57 Ibid,Oesch, 2016

58 Ibid,Oesch, 2016

59 Kerr Philip, Editor, The Penguin Book of Lies, Penguin Books, 1990, P.3

60 Darwin Charles, The Descent of Man, Wordsworth Classics of World Literature, Limitededition 2013, Chapter 2, PP. 43-44

61 Coleridge, S.T., The Compete Poems, edited by Willima Keach, Penguin Books, 1997, pp.138-139

62 Aristotle, Poetics, trans.Kenneth McLeish, Nick Hern Books, London;2008,p.3

63 Ibid.Poetics, P. 9

64 For details of Belief-dependent Realism and Model-dependent Realism see the book The Grand Design 2010 by late cosmologist Stephen Hawking and mathematician Leonard Mlodinow

65 Plato, Republic, translated by John L. Davies and David j. Vaughan, Wordsworth Classics of World Literature, 1997, PP. 58-59

66 Ibid., Republic, PP 66-67

67 Ibid., Republic, P.327

68 Ibid., Republic, P. 216 (506a-e)

69 Aristotle's Posterior Analytics, translated by Edmund Spencer Bouchier, Pub. B.H Blackwell,Oxford 1901, Book I, Chapter xix

70 Ibid., Posterior analytics, Book I, Chapter xvII

71 Aristotle, De Anima, Trans.by Hugh Lawson-Tangred, Penguin Books, 1986 pp.198-199 427b, 428a,

72 The Ethics of Aristotle , The Nicomachean ethics, trans.,J.A.K Thomson,Penguin Books1976,p.317

73 Aristotle The Nicomachean Ethics, Wordsworth Classics of World Literature, trans. Harris Rackham,1996,p.269

74 Ibid., Aristotle, P.268

75 Machiavelli Niccolo, The Prince, translated by George Bull, Penguin Classics, 1981, P.101

76 Ibid., Machiavelli, P. 101

77 Ibid., Machiavelli, P. 100

78 Ibid., Machiavelli, P. 100

79 Saint Augustine, Lying, from Treatises on Various subjects (The Fathers of the

Church, Vol. 16, translated by Sister Mary Sarah Muldowney, Published by Catholic University of America, 1952,P. 55

80 Ibid., St Augustine, P.55

81 Ibid St. Augustine, 67-68

82 Ibid., St. Augustine, P.57

83 Ibid; St. Augustine, P.54

84 Michel de Montaine, Essays, Translated by J. M. Cohen, Penguin Books, 1963, P.11

85 Ibid, Montaine, P.11

86 Poulshock J. W. (2006), Language and Morality: Evolution, Altruism, and Linguistic Moral Mechanisms, University of Edinburgh, P.1

87 Heaton, J. & Groves, J. Introducing Wittgenstein, Icon Books, UK., 2003, P.119

88 Hoopoe bird is a colourful bird found across Afro-Eurasia, notable for its distinct crown of feathers Google search

89 Bentham, J., An Introduction to the Principles of Morals and Legislation, Edited by J. H. Burns and H. L. A. Hart, 1970, Athlone P. London, P.11

90 Delapp, K., and Henkel, J., Ed., Lying and Truthfulness, 2016, Hackett Publishing, P. 84

91 Ordination is a term in ecology which means a statistical technique in which data from a large number of sites or populations are represented in a multidimensional space.

92 Rousseau, J. J. The Social Contract and Discourse, Trans. By D. H. Cole, Pub. J. M. Dent & Sons Ltd., London, 1975, P. 66

93 Ibid., Rousseau, P.168

94 Ibid., Rousseau, P.65

95 Ibid., Rousseau, P. 64

96 Ibid., Rousseau, P. 65

97 Hobbes, T., Leviathan, ed., C. B. Macpherson, Penguin Books, 1988, P.213

98 Ibid., Hobbes, P. 122

99 Ibid., Hobbes, P., 215

100 Ibid., Hobbes, P. 214

101 Ibid., Hobbes, P. 180

102 Ibid., Hobbes, P. 131

103 Cahn, S. M. Classics of Political and Moral Philosophy, Oxford University Press, 2002, P. 572

104 Ibid., Cahn, P. 572

105 Selby-Bigge, L.A. ed., Enquiry Concerning Human Understanding,3rd edn., Oxford University Press, 1975, P. 83

106 Locke, J., An Essay Concerning Human Understanding, Collins,The Fontana Library, 1964, P. 89

References

- Abeler J.A. Becker A. Falk A., (2012), Truth-telling: a representative assessment. Inst. Study Labor 6919 1-18
- Ackrill, J. L. (1990), Aristotle the Philosopher, Clarendon Press, Oxford
- Aktipis, A & Maley C. C., (2017) Cooperation and Cheating as Innovation: insights from cellular societies, https://doi.org/10.1098/rstb.2016.0421
- Alcamo, E. (1997), Anatomy Colouring Book, Random House, Inc. New York
- Anscombe E. G. (1981), Ethics, Religion and Politics, Collected Philosophical Papers, Vol. III, Wiley Blackwell Publishing
- Aristotle, (2008), Poetics, Trans. Kenneth McLeish, Nick Hern Books, London
- Aristotle, (1996), The Nicomachean Ethics, trans., Harris Rackham, Wordsworth Classics
- Aristotle, (1986), De Anima – On the Soul, trans., Hugh Lawson-Tancred, Penguin Books
- Aristotle, (1995), Politics, trans., Ernest Barker, Oxford University Press
- Aristotle, (1976), Ethics, trans., Hugh Tredennick, Penguin Books
- Aristotle, (1901), Posterior Analytics, trans., by Edmund Spencer Boucher, Pub., by B. H. Blackwell, Oxford
- Ashworth, P. (2000), Psychology and Human Nature, Psychology Press
- Atkinson, N. (2004), Darwin Meets Chomsky, Charles Darwin spotted it, https://www.the-scientist.com > ...
- Badore, M. (2014), Cuckoos and Crows Teach Us How Parasites can be Good, University of Oviedo, Spain, printed in 'Science Magazine', March 20, edition
- Barash, D. (1981), Sociobiology: The Whispering Within, Fontana/Collins

- Barnes, J. (1982), Aristotle, Oxford University Press
- Barrett, P. et al eds. (1987), Charles Darwin's Notebooks, 1836-1841, Cambridge University Press and British Museum
- Bateson, G. (1985), Steps To An Ecology of Mind, Ballantine, New York
- Behe, M.J. (1998), Darwin's Black Box, Simon & Schuster
- Bel, L. (2019), The Neuroscience of Lying, https://brainworldmagazine.com > ...
- Benner, E. (2017), Be Like The Fox: Machiavelli's Lifelong Quest for Freedom, Allen Lane
- Bentham, J. (1970), An Introduction to the Principles of Morals and Legislation, edit. By J. H. Burns & H. L. A. Hart, Athlone P. London
- Bergman, J. (2010), Freud and Darwinism, Journal of Creation 24 (2), Retrieved from https://creation.com>pdfs
- Berndt, T. J. (1992), Child Development, Harcourt Brace Jovanovich College Publishers
- Bloom, P. (2010), How Pleasure Works, Why We Like, What We Like, Vintage Books
- Bok, S. (1979), Lying: Moral Choice in Public and Private Life, Vintage Books
- Brettler, M. Z. (2015), Basing Judaism on Truth: Does the Torah Lie? https://www.thetorah.com , blogs
- Byrne, R. (1995), The Thinking Ape, Oxford University Press
- Cahn, S. M. (2002), Classics of Political and Moral Philosophy, Oxford University Press
- Carlson, N. R. (2007), Physiology of Behavior, Ninth Ed., Pearson
- Cartwright, J. (2000), Evolution and Human Behavior, Palgrave

- Carver, C. S. & Scheler, M. J. (2004), Perspectives on Personality, 5th edit., Pearson, Boston
- Cherry, K. (2019), Freud's Theory of the Id in Psychology, https://www.verywellmind.com > w...
- Chomsky, N. (2002), On Nature and Language, Cambridge University press
- Clutton-Brock, T. H. & Harvey P. H. Editors, (1978), Readings in Sociology, W. H. Freeman And Company
- Collodi, C. (2002), Pinocchio, trans., M. A. Murray, Penguin Books
- Comey, J. (2018), A Higher Loyalty, Truths, Lies, and Leadership, MacMillan
- Connell, W. (2018), The inverted advice of Niccolo Machiavelli, https://mobile.twittwer.com > status
- Corn, D., (2003), The Lies of George W. Bush, Crown Publishers, USA
- Darwin, C. (1985), The Origin of Species, Penguin Classic
- Darwin, C. (2013), The Descent of Man, Wordsworth Classics of World Literature, Limited Edition
- Dayantis, H. (2016), How Lying Takes Our Brains Down a 'Slippery Slope' UCL research, http://neurosciencenews.com/lying-emotion-psychology-5345/.
- Delapp, K, & Henkel, J. editors, (2016), Lying and Truthfulness, Hackett Publishing
- Dews, P. Edit., (1999), Habermas A Critical Reader, Blackwell Publishers
- Dickinson G. L. (2003), Plato and his Dialogues, University Press of the Pacific Honolulu
- Doody, R. (2018), Lying and Denying, scholar.google.co.il,citations
- Dor, D. (2017), the Role of the Lie in the Evolution of Human Language,

https://www.sciencedirect.com , pii

- Dougherty, E. (2011), What are thoughts made of? https://engineering.mit.edu , engage
- Duncan, R. & Weston-Smith, M. (1979), Lying Truths, A Critical Scrutiny of Current Beliefs and Conventions, Pergamon Press
- Evans, D. and Zarate, O. (1999), Introducing Evolutionary Psychology, Icon Books UK, Totem Books USA
- Farber, M. (1968), Basic Issues of Philosophy, Experience, Reality and Human Values, Harper Torchbooks
- Fitch, W. T. (2010), The Evolution of Language, Cambridge University Press
- Franken, A. (2004), Lies and the Lying Liars Who tell them, Penguin Books
- Freud, S. (1991), Introductory Lectures on Psychoanalysis, Penguin Books
- Freud, S. Beyond the Pleasure Principle, The International Psycho-Analytical library, No 4, Trans. By Caroline Jane Mary Hubback from the 2nd German Edit., Revised and edit., by Ernest Jones, the International Psycho-Analytical Press, London
- Galliott, J. & Reed, W., Edit., (2016), Ethics and the Future of Spying, Routledge
- Gerasimo, P. (2003), Emotional Biochemistry, https://experiencelife.com , article
- Gleason, J. B. (1985), The Development of Language, Charles E. Merrill Pu. Company, Ohio, USA
- Goleman, D. (1998), Vital Lies, Simple Truths, The Psychology of Self Deception, Bloomsbury
- Greene, J. D. & Paxton, J. M. (2009), Patterns of neural activity associated with honest and dishonest moral decisions. Proc. Nat. Acad. Sci. U.S.A 106, 1-6. Doi:10.1073/pnas.0900152106

- Gribbin, M. & J. (1998), Being Human, Putting People in an Evolutionary Perspective, Phoenix
- Gunderman, R. (2016), Why you shouldn't blame lying on the brain, https://theconversation.com > w...
- Hale, J. R. (1972), Machiavelli and Renaissance Italy, Penguin Books
- Hall, C. S. & Gardner, L. (1978), Theories of Personality, 3rd edit., John Wiley & Sons, New York
- Hartley, G. & Karinch M. (2009), How to Spot a Liar, Why people don't tell the truth and how you can catch them, Logoprint, SpA, Trento
- Hawkins, S. & Mlodinow, L. (2010), Belief Dependent Realism and Model-dependent Realism in The Grand Design, Bantam Books
- Heaton, J. & Groves, J. (2003), Introducing Wittgenstein, Icon Books, Uk.
- Heyes, C. (1994), Social Cognition in Primates, edit., N. J. Mackintosh, printed in Animal Learning and Cognition, Academic Press
- Hill, S. edit. & trans., (2004), The Power of Plato, An Anthology of New Translations, Duckworth
- Hobbes, T. (1988), Leviathan, ed. By C. B. Macpherson, Penguin Books
- Holiday, R. (2012), Trust Me I'm Lying, Confessions of a Media Manipulator, Penguin Books Ltd.
- Homo Erectus: Methods of Communication-UWC (Planet.uwc. ac.za>loe>page-314)
- Hooper, J. & Kraczyna, A. (2019), The Truth about Pinocchio's Nose, https://en.wikipedia.org , wiki , Pin...
- Hopper, S. D. & Lambers, H. (2009), Darwin as a Plant Scientist: A Southern Hemisphere Perspective, published in Trends in Plant Science Vol. XXX, No. X 2009
- Jiang, W., et al. (2015), Decoding the processing of lying using functional connectivity MRI, https://www.ncbi.nih.gov > pmc

- Johnson, G. (2015), Cellular Cheaters Give Rise to Cancer, retrieved from New York Times, https://www.nytimes.science28jul2015
- Jones, A. (2018), Flora fatale: The Carnivorous Plant that Scared Charles Darwin, Google Search
- Jowett, B. (2000), Selected Dialogues of Plato, translated by Benjamin Jowett, The Modern Library NY
- Kaplan, D. & Manners, R. A. (1972), Cultural Theory, Prentice-Hall, Inc. New Jersey
- Karim, A. et al. (2010), The Truth about Lying: Inhibition of the Anterior Prefrontal Cortex Improves Deceptive Behavior, https://doi.org/10.1093/cercor/bhp090
- Kaufmann, W. (1976), The Portable Nietzsche, Penguin Books
- Kenrick, D. T. and Luce, C. L. (2004), The Functional Mind, Readings in Evolutionary Psychology, Pearson
- Kern, E. Edit., (1962), Sartre, A Collection of Critical Essays, Prentice-Hall
- Kerr, P. (1990), The Penguin Book of Lies, Penguin Books
- Key West, Florida Forum, (2008), How Our Brains are Wired for Belief, https://www.pewforum.org > how-o...
- King, B. The Lying Ape (2007), An Honest Guide To The World Of Deception, Icon Books
- Kleinman, P. (2013), Philosophy 101, Adams Media
- Knight, C. (1998), 'Ritual/speech coevolution: a solution to the problem of deception,' from Approaches to the Evolution of Language, edited by Hurford J. R., Studdert-Kennedy M., & Knight C., Cambridge University Press
- Kornet, A. (1997), The Truth about Lying, https://www.psychologytoday.com , ...
- Korsgaard C. M. (1986), The Right to Lie: Kant on Dealing with Evil,

https://dash.harvard.edu , ko...

- Kottak, C. P. (1974) The Exploration of Human Diversity, Random House, New York

- Kotulak, R. (2004), Lips can lie, but your brain will spill the beans, https://www.chicagotribune.com > ...

- Krebs, J. R. & Davies N. B., Editors, (1984), Behavoural Ecology: An Evolutionary Approach, Oxford University Press

- Krebs, J. R. & Dawkins R., Animal Signals: Mind-reading and Manipulation, retrieved from Behavoural Ecology, An Evolutionary Approach, 2nd ed., edited by Krebs and Davies, Blackwell Scientific Publications, Oxford

- Kucich, J. (1994) The Power Of Lies, Transgression in Victorian Fiction, Cornell University Press

- Liberman, M. (2009), Musical protolanguage: Darwin's theory of language evolution revisited, under Language and music, https://languagelog.ldc.upenn.edu , ...

- Lincoln, C. (2015), How to Stop Lying, The Ultimate Cure Guide for Pathological Liars and Compulsive Liars, Amazon Distribution

- Livingstone-Smith, D. (2004), Why We Lie, St. Martin's Press, N.Y.

- Livingston Smith, D. (1991), Hidden Conversations, An Introduction to Communicative Psychoanalysis, Tavistock/Routledge

- Locke, J. (1964), An Essay Concerning Human Understanding, Collins, The Fontana Library

- Lowes Dickinson, G. (2003), Plato and his Dialogues, University Press of the Pacific

- Lucas, E. (2012), Deception: Spies, Lies and How Russia Dupes the West, Bloomsbury Publishing

- Lutzer, E. W. (2009), 10 lies about God, And The Truths That Shatter Deception, Kregel Publishers

- Machiavelli, N. (1985), The Prince, trans. By George Bull, Penguin Books
- Machiavelli, N. (2001), Power: Get it, Use it, Keep it
- Machiavelli, N. (2010), On Conspiracies, trans., Leslie J. and Walker SJ, penguin Books
- Macknik, S. L., Martinez-Conde, S. & Blakeslee, S. (2011), Sleights of Mind, What Neuroscience of Magic Reveals about our Brains, Profile Books
- Maclay, K. (2014), Study links honesty to prefrontal region of the brain, https://news.berkeley.edu > study-lin...
- Macnabb, D. G. C. Edit., (1962), David Hume: A Treatise of Human Nature, The Fontana Library
- Magee, B. (1998), The Great Philosophers, An Introduction to Western Philosophy, Oxford University Press
- Mahon J. E. (2009), The Truth about Kant On Lies, https://philarchive.org , MAHTTA
- Malim, T., Birch, A. and Hayward, S. (1966), Comparative Psychology, Human and Animal Behaviour: A Sociobiological Approach
- Martin Alcoff, L. Edit., (2007), Epistemology: the Big Questions, Blackwell Publishing
- Mayfield Medicine, Anatomy of the Brain, https://mayfieldmedicine.com , pe-anat...
- McLeod, S. (2018), The Preoperational Stage of Cognitive Development, https://www.simplypsychology.org
- McLeod, S. (2019), Id, Ego and Superego, https://english-online.fi > materials
- McLeod, S. (2019), Freud's Psychosexual Stages of Development, https://www.simplypsychology.org > ...
- Mark J. J. (2019) Plato's Lie In The Soul, https://www.ancient.eu > article

> pl...

- Martin, C. (2015), Love and Lies, Harvill Secker, London
- Melling, D. J. (1990), Understanding Plato, Oxford University Press
- Millholand, L. & Srikanthan, S. (2014), Area of the brain responsible for honesty identified, www.collegiatetimes.com > news > a...
- Millikan, R. (1984), Language, Thought and Other Biological Categories, Cambridge, MIT press
- Montaigne, M. (1963), Essays, Penguin Classics
- Moore, K. L. and Dalley, A. F. (1999), Clinical Oriented Anatomy, Fourth Ed. Lippincott Williams & Wilkins
- Morrissey, C. (2019), The Truth about Plato's "Noble Lie", https://theimaginativeconservative.org > ...
- Moss, J and Schwab, W. (2019), The Birth of Belief – PhilPapers, https://philpapers.org > MOSTBO-2
- Muller M. (1872) On Darwin's Philosophy of Language, https://www.nature.com , news
- Nanva, A. (2019), Mythologies of Late, Homage to Roland Barthes, Austin Macauley Publishers
- Newman, M. L. et al (2016), Lying Words: Predicting Deception from Linguistic Styles, https://journals.sagepub.com , abs
- Nietzsche F. (1971), Thus Spoke Zarathustra, trans., R. J. Hollingdale, Penguin Books
- Nietzsche F. (1990), Beyond Good and Evil, trans. R. J. Hollingdale, Penguin Books
- Nietzsche F. (1956), The Birth of Tragedy and The Genealogy of Morals, trans., Francis Golffing, Doubleday Anchor Books
- Oborne, P. (2005), The Rise of Political Lying, Cox & Wyman Ltd. GB.

- Oesch N. (2016), Deception as a Derived Function of Language, https://www.ncbi.nlm.nih.gov , pub...

- Ofen, N., et al, (2017), Neural correlates of deception: lying about past events and personal beliefs, https://www.ncbi.nlm.gov > pub...

- Özkaya et al. (2018), Cheating on Cheaters stabilizes Cooperation in Pseudomonas aeruginosa, Current Biology, https://doi.org/10.1016/j.cub.2018.04.093

- Parker, J. (2016), Myrmecophily in Beetles (Coleoptera): Evolutionary Patterns and Biological Mechanisms, ret. From http://www.antwiki.org/wiki/images/5/52/parker-2016.pdf

- Peplow, M. (2004), Brain imaging could spot liars, https://www.nature.com > news > full

- Piaget, J. (1929), The Child's Conception of the World, trans., by Andrew Tomlinson, Harcourt, Brace & World Inc, New York

- Pinker, S. (1994), The Language Instinct, Penguin Books

- Pinker, S. (1997), How the Mind Works, Norton

- Pinker, S. and Bloom, P. (1990), Natural language and natural selection, Behav.Brain Sci. 13,707-784. Doi:10.1017/S0140525X00081061

- Plato, (1997), Republic, trans., John Llewelyn Davies & David James Vaughan, Wordsworth Classics of World Literature

- Plato, (1987), Theaetetus, trans., Robin A. H. Waterfield, Penguin Books

- Plato, (1951), The Symposium, trans., Walter Hamilton Penguin Books

- Poulshock, J. W. (2006), Language and Morality: Evolution, Altruism, and Linguistic Moral Mechanisms, University of Edinburgh

- Proverbio, A.M. Vanuttelli, M.E. and Adorni, R. (2013), Can You Catch a Liar? How Negative Emotions Affect Brain Responses When Lying or Telling the Truth, https://doi.org/10.137l/journal.pone.0059383

- Pruss A. R. (2001), Lying, Deception and Kant, alexanderpruss.com , papers , Lying…

- Radick G. (2002), Darwin on Language and Selection, https://pdfs.semanticscholar.org
- Rolston III, H. (1999), Genes, Genesis and God, Values and Their Origins in Natural and Human History, Cambridge University Press
- Rosen, S. (1974), G. W. F. Hegel, An Introduction to the Science of Wisdom, Yale University Press
- Rouse, W. H. D. trans. (1963), Great Dialogues of Plato, The New American Library
- Rousseau, J. J. (1975), The Social Contract and Discourses, trans. G. D. H. Cole, pub. J. M. Dent & Sons Ltd., London
- Rubner. A. (2006), The Mendacious Colours of Democracy, imprint-academic.com
- Saint Augustine, Lying, translated by sister Mary Sarah Muldowney (2018), Published By Catholic University of America Press, http://www.jstor.org/stable/j.ctt32b2mf.5
- Saint Augustine, (1887) On Lying, translated by Rev. H. Browne, Corpus Christi College, Cambridge
- Saint Thomas Aquinas, Summa Theologica 1265-1274, Vices Opposed to Truth, www.godrules.net , library , Aquinas
- Salman A. & Parens H. editors, (2009), Lying, Cheating, and Carrying On: Developmental, Clinical and Sociocultural Aspects of Dishonesty and Deceit, Jason Aronson
- Sanders, L. (2016), Lying sets up a liar's brain to lie more, https://www.sciencenewsforstudents.org > …
- Sartre, J. P., (1997), Existentialism and Human Emotions, trans., Hazel E. Barnes, Carol Publishing Group
- Sartre J. P. (1989), Being and Nothingness, Trans., Hazel e. Barnes, Routledge

- Schaarschmidt, T. (2018) The Art of Lying, www.scientficamerican.com
- Scott-Phillips, T. C. (2010), Evolutionary Psychology and the Origins of Language, https://en.wikipedia.org , wiki , Ev..., Editorial for the special issue of Journal of Evolutionary Psychology on the Evolution of Language
- Searcy, W. A. & Nowicki, S. (2007), The Evolution of Animal Communication: Reliability and Deception in Signalling Systems, Princeton University Press
- Selby-Bigge, L. A. Editor, (1975), Enquiry Concerning Human Understanding, 3rd ed., Oxford University Press
- Shermer, M. (2011), The Believing Brain, From Spiritual Faiths to Political Convictions, Robinson
- Shettleworth, S. J. (1998), Cognition, Evolution and Behavior, Oxford University Press
- Sick, H., Nova Frontiera, retrieved from https://www.wikiaves.com>forum
- Southwell, D. (2005), Secrets and Lies, Exposing The World Of Cover-ups And Deception, Carlton Books
- Steffens, M. (2007), Natural Born Liars, www.abs.net.au
- Stevenson, L. (1987), Seven Theories of Human Nature, 2nd edit., Oxford University Press
- Stevenson, L. edt., (2000), The Study of Human Nature, A Reader, Oxford University Press
- Stone, A. (2018), Is Your Child Lying to You? That's Good, https://www.nytimes.com > chil...
- Stove, D. (1995), Darwinian Fairytales, Encounter Books
- Strom B. D. (2018), St. Augustine On Lying from Reflections on Theology and Moral Philosophy, https://www.seekingvityueandwisdom.com...

- Sutliff, U. (2005), First Evidence Of Brain Abnormalities Found In Pathological Liars, University of Southern California, https://pressroom.usc.edu > first-evi...
- Sutliff, U. (2005), Liar's Brains Wired Differently, https://news.usc.edu> Liars-Brains-...
- Talwar, V. and Lee K. (2008), Social and Cognitive Correlates of Children's Lying Behavior, https://www.ncbi.nlm.nih.gov , pub...
- Thibodeaux, W. (2017), Here's What Happens in Your Brain When You Repeatedly Lie, https://www.inc.com > this-brain-stu...
- Thomson, J. J. & Dworkin, G. Trans., (1968), Ethics, Harper & Row Publishers
- Tomasello, M. (2008), Origins of Human Communication, Cambridge, MIT Press
- Tortora, G. J. & Anagnostakos, N. P. (1987), Principles of Anatomy and Physiology, 5th edition, Harper & Row Publishers
- Tuckman, B. W. (1992), Educational Psychology, From Theory to Application, Harcourt Brace Jovanovich College Publishers
- Twain, M. (2016), The Stolen White Elephant, Penguin Classics
- Twain, M. (2013), The Wit and Wisdom of Mark Twain, A Book of Quotations, Dover Publications, INC
- University of Cambridge, Darwin Correspondence Project, The Origin of Language.jpg
- Varden H. (2010), Kant and Lying to the Murderer at the Door..., onlinelibrary.wiley.com
- Viroli, M. (1998), Machiavelli, Oxford University Press
- Vitelli, R. (2013), When does Lying Begin? https://www.psychologytoday.com
- Waldron, T. P. C (1985), Principles of Language & Mind, Routledge &

Kegan Paul

- Walters, S. B. (2000), The Truth about Lying, Sourcebooks, INC, Illinois, USA
- Ward, J. (2012), The Student's Guide to Social Neuroscience, Psychology Press
- Warmelink, L. (2014), Children lie from the age of two, so here's how to get them to tell the truth, https://theconversation.com , pants-...
- Webster, B. (1985), Guile and Deception: The Evolution of Animal Courtship, Section C, P.3, The New York Times, reprint www.nytimes.com/1985/2/science/guile-and-deception-the-evolution-of-animal-courtship.html
- Wilde, O. (2003), The Decay of Lying and Other Essays, Penguin Classics
- Wood, A. W. (2011), Kant and the right to lie reviewed essay: On a supposed right to lie from philanthropy, by Immanuel Kant (1797), Stanford University, Faculty of Humanities
- Xu F. et al (2010), Lying and Truth-Telling in Children: From Concept to Action, https://www.ncbi.nlm.nih.gov , pub...

* * *

The images used in the book are from the following Pixabay authors:

- Front cover and on page 17 by Gordon Johnson.
- Page 6, originally from OpenClipart-Vectors.
- Page 12, by Mohamed Hassan.
- Page 14, from OpenClipart-Vectors.
- Page 62, originally from Cdd20.
- Page 80, originally from Merio.

- Page 96, from OpenClipart-Vectors.
- Page 129, originally from John Hain.

Lightning Source UK Ltd.
Milton Keynes UK
UKHW021225171220
375421UK00008B/436